THE GOLDEN ORIOLE

THE GOLDEN ORIOLE

PAUL MASON
JAKE ALLSOP

T & AD POYSER
LONDON

Published 2009 by T & AD Poyser, an imprint of A&C Black Publishers Ltd.,
36 Soho Square, London W1D 3QY

www.acblack.com

ISBN 978–0–7136–7683–9

A CIP catalogue record for this book is available from the British Library

Commissioning Editor: Nigel Redman
Project Editor: Jim Martin

Design: J&L Composition Ltd, Scarborough, North Yorkshire

Printed and bound in Great Britain

FSC www.fsc.org MIX Paper from responsible sources FSC® C013604

Contents

Acknowledgements

Thanks go to the many knowledgeable farm and estate workers who alert us every season to the arrival or presence of orioles, and to the farmers and landowners who give us access to their property to carry out our work. Many individuals have helped with fieldwork over the years, and we apologise that there are too many to list individually.

We thank all the people who have patiently responded to our many queries and requests for information, including Mark Cocker, David Lingard (LIPU-UK), Gerald Rolfe, Martin Scott (Scilly Isles) and Richard Thomas. Special mention should be made of the following UK contacts – Chris du Feu and Euring for supplying most of the ringing data used, and Bob Scott for his help with Bulgarian contacts.

We would also like to thank Richard Thomas (formerly of BirdLife International) for his unstinting help in making contact with the many overseas people listed below; Roger Riddington, editor of *British Birds*, for similar help in finding useful contacts in other countries; and Mike Jennings, for sharing his considerable knowledge of Arabian birds; Peter Jones and Bruno Walther for supplying their text on the *Oriolus* genus for inclusion in the forthcoming volume of *Handbook of Birds of the World.*

Among our many overseas correspondents, we acknowledge (listed alphabetically by country): Tom Talbot (Australia), Hans-Martin Berg (Austria), Michael Dvorak (Austria), Elchin Sultanov (Azerbaijan), Wouter Favets (Belgium), Glenn Vermeersch (Belgium), Boris Barov (Bulgaria), Petar Tankov (Bulgaria), Professor Ma Ming (Xingjiang, China), Jasmina Muzinic (Croatia), Karel Stastny (Czech Republic), Rolf Cristiansen (Denmark), Gert Fahlberg. (Denmark), Mary Magali (Egypt), Margus Ellermaa (Estonia), Tom Lindroos (Finland), Zura Javakhishvili (Georgia), Jochen Dierschke (Germany), Sabine Baumann (Germany), Martin Flade (Germany), Triantaphyllos Akriotis (Greece), Maria Dimaki (Greece), Arnoud van den Berg (Holland), Akos Moniki (Hungary), Yoav Perlmann (Israel), Marco Gustin (LIPU, Italy), Jamshid Mansoori (Iran), Ian Andrews (Jordan), Agnese Balandioa (Latvia), Ghassan Ramadan-Jaradi (Lebanon), Joseph Mangion (Malta), Branco Micevski (Macedonia), Ludwik Tomiałojc (Poland), Tadeusz Stawarczyk (Poland), Helder George da Silva Costa (Portugal), Ray Tipper (Portugal), Moldovan Istvan (Romania/Egypt), Attila Sandor (Romania), Jeff Gordon (southern Russia), Igor Fefelov (Irkutsk, Russia), Sergey Gashkov (Siberia, Russia), Rudolf Kropil (Slovakia), Andrej Figelj (Slovenia), Dan Hammerlund (Sweden), Gianlucca Serra (Syria), Geoff Welch (Turkey).

We are especially grateful to the following overseas colleagues: Pascal Etienne (France) for supplying his papers, for showing us round his oriole breeding sites in northern France and for entertaining us royally; Walter Hovorka (Austria) for freely providing information from his own researches; Jevgeni Shergalin (Estonia), for researching and translating the many Russian papers used in the text; Sabine Baumann (Germany) for generously sharing the results of her research (including her studies of oriole songs and calls in an as-yet unpublished paper), and for being willing to discuss many points of detail; Marco Dragonetti (Italy), for helping us to interpret, create and translate sonograms. He too gave freely of his time to help in various other ways; Rob Bijlsma (Holland), for contributing a great deal in addition to making his papers available to us, and for kindly translating the Dutch folk song which appears in the Introduction; Igor Fefelov (Siberia) for the many email exchanges in answer to our numerous questions about the complexity of oriole distribution, breeding and movement in Siberia; Valery Moseykin (Russia) for his invaluable descriptions of his observations in Russia, and for permission to use his striking photographs in this book.

We give special thanks to Chris Knights for the considerable help he gave photographing and filming orioles at the nest, and for his kind permission to allow some of his photographs to be included in the book. He and GOG member Malcolm Rains worked together at most sites; many of Malcolm's pictures are also used in the book. We also thank Alastair Burn for his drawings, which so enhance the book. Howard Slatter for his prompt and invaluable help with statistical issues; Peter Fraser and the late Michael Rogers, who produce the annual report on scarce migrant birds in Britain in *British Birds* journal, regarding migration; and John Sherwood and Christine Alder, librarians at BirdLife International. We are grateful for help of various kinds that we received from the RSPB, including permission to adapt data on nest orientation used in the pie charts and in particular thank Darren Oakley-Martin of the RSPB for help in tracking down references. The Forestry Commission, and in particular the late John Jobling, helped us to identify poplar cultivars and Arnold Beaton, also of the Forestry Commission until his retirement, also helped with these trees. We thank the Forestry Authority for permission to use information from Research Information Note 265 regarding characteristics of poplar cultivars.

At the stage of the first draft, we were fortunate to have two experts read through it – Peter Wilkinson, formerly of BirdLife International, and Tony Prater, formerly of the RSPB. Thanks to them, we were able to eliminate many errors and omissions. It is a better book for their help, but we take responsibility for any errors that remain.

On a personal note, Paul would like to thank his patient wife, Cicely, for allowing him to spend hours at the computer when he should have paid her more attention. It is a great pity she did not live to see the end product, because although she was no birdwatcher, she did love to see and hear Golden Orioles.

This book, like all of its kind, cannot be regarded as the last word on the subject. It is merely a snapshot of our knowledge at the time of writing, with undoubtedly more discoveries to be made. Will the splitting of the Indian Golden Oriole, *Oriolus kundoo*, for example, be confirmed or refuted by DNA analysis, or by evidence that it interbreeds with *Oriolus oriolus*?

Paul Mason and Jake Allsop
March 2009

Introduction

The Golden Oriole, as most oriole watchers who study the bird will surely agree, is the most frustrating, intelligent, beautiful, characterful, acrobatic, brave, diverse, successful and exciting arboreal passerine there is.

It is perhaps the most widespread arboreal migrant in the Palaearctic region, breeding from Portugal (9° W) in the west to near Lake Baikal (110° E) and Mongolia in the east. It breeds from the Saudi Arabian Desert (25° S) in the south to Karelia in Finland and Russia (63° N) in the north.

It nests in the most diverse habitats, from dry sandy areas to intensive agriculture and damp swampy forests, from commercial plantations of regimented trees to the depths of primeval forests. The Golden Oriole can be found breeding from below sea level to over 1,500m above sea level and climbs to over 2,500m when on migration. Its nests are placed in bushes and trees at heights from 1.5m to 30m.

It eats a most diverse range of invertebrates, from minuscule beetles to large bush crickets, locusts, large hairy caterpillars and worms, and all manner of fruits up to the size of figs and mangoes, which it collects anywhere from the ground to the highest treetop. What's more, the Golden Oriole also drinks nectar. It is also arguably the longest-distance arboreal migrant, with some travelling a maximum of 13,000km if flying in a straight line (which they probably don't). The only longer-distance passerine migrant is the Northern Wheatear, which abhors trees.

To some the Golden Oriole is regarded as a great aid in eliminating pest infestations, yet to others it is in itself a pest.

Its voice ranges from two strident calls to a ventriloquial, flute-like song with a mixture of up to 50 different song phrases or types.

It is popularly regarded as among the most difficult of birds to see, spending most of its time hidden frustratingly in the dappling effect of the trembling leaves of poplar canopies. It acts like a magnet for twitchers and general public alike, provoking the typical commentary, 'Golden Oriole where, where. I can't see it. Where is it? Gosh, isn't it beautiful. What a lovely call. Oh it's gone. Drat.'

It's a bird that needs time but, when given it, the Golden Oriole gives all a thrill and a satisfied mind.

The following traditional children's poem from Holland has been freely translated by Rob Bijlsma, who also says the first line literally translates as 'Let's get out' but 'Going up the country' is very much better as it covers the feeling of the moment and is rhythmically helpful for singing this children's song.

Kom mee naar buiten allemaal,
Dan zoeken wij de wielewaal—
En horen wij die muzikant,
Dan is zomer weer in 't land!
Dudeldjo klinkt zijn lied,
Dudeldjo klinkt zijn lied,
Dudeldjo en anders niet.
Hij woont in 't dichte eikenbos—
Gekleed in gouden vederdos.
Daar jodelt hij op zijn schalmei—
Tovert onze harten blij.
Dudeldjo klinkt zijn lied,
Dudeldjo klinkt zijn lied,
Dudeldjo en anders niet.

Going up the country,
Looking for the oriole—
Summer has arrived again,
When calling starts this songster.
Dudeljo resounds his song
Dudeljo resounds his song
Dudeljo and nothing more.
He lives in oakwood dense—
Clad in golden dress.
Sparks our hearts with happiness,
When yodelling his shawm.
Dudeljo resounds his song
Dudeljo resounds his song
Dudeljo and nothing more.

He found only one other poem about Golden Orioles in Dutch, even though many poets there are involved in light verse on nature. This poem is untranslatable, but speaks for itself, being mainly based on the word sounds, i.e. it is largely onomatopoeic. It is by Hans Warren from his book *Betreffende vogels*, 'Regarding Birds', published in 1974 by Erven Thomas Rap, Amsterdam.

De Wielewaal

Oriolus, Gele gouw, Gele wiewouw,
Loriot, Figo l'Aouriaou, Migliora,
Ajulu, Gabrieli, Agruppa filu,
Rigogolo, Crusuelo, Pirol,
Schulz von Milo, Schulz von Bülow,
Oropendola. 'Hio bulo!
Gidleo gitatidlio gigilio
gipliagiblio gidleaah!

THE BACKGROUND TO THE GOLDEN ORIOLE GROUP

The Golden Oriole Group was formed in 1986 by several birders who had individually been watching Golden Orioles at a 350-ha hybrid poplar plantation, planted from about 1954 onwards by the famous matchmakers Bryant & May Ltd on Shepherds Farm, Lakenheath, Suffolk. At that stage the company wished to ensure a constant British supply of poplar wood to make their matches, in order to reduce their dependency on imports. By 1980, however, they had changed course, after a more liberal free-trade policy became the political norm. It was found that it was cheaper to import poplar wood rather than continue a 'home-grown' programme, and it was suggested that foreign-grown trees were of better quality. This also coincided with a decline in the use of matches generally. No doubt with a desire to realise capital assets the estate was sold back to a farming company, whereupon the trees were gradually felled and the fields returned to agricultural crops. Whilst all this was going on Golden Orioles, hitherto only rare, spasmodic breeding birds in Britain, had found that the environment that had been constructed was much to their liking, and the population had increased from one pair in 1965 to a possible 14 pairs by 1980.

On learning of the proposals for the reduction in the area of poplars, the concerned birders agreed to form the Golden Oriole Group (GOG), to monitor cooperatively what happened to the fortunes of the birds. Would they disappear altogether or would the birds disperse to other known plantations in the surrounding fenland? It was found that the latter was the case, so a programme of study was instituted. The GOG has since established the type and size of habitat best-suited to orioles, and has establihsed the hybrid cultivars of poplar that the birds prefer to use for nesting. It has also studied their behaviour through intensive observation, and their diet by photography and filming at nests. Grant aid has enabled researchers to produce a scientific report listing invertebrates available in poplar as possible food for nesting birds. As many birds as possible have been ringed and their breeding biology has been fully documented, as have numbers breeding in any one year in the study area. The information on population size is sent to the Rare Breeding Birds Panel, the RSPB and Natural England. In May, June and July of 1987 the RSPB installed a researcher, Jez Dagley, on the estate; Dagley wrote a confidential report for the RSPB and the GOG, which has been the basis for further investigations. In 1995 a national survey was undertaken to ascertain as best it could how far orioles travelled throughout mainland Britain. The GOG has been able to advise farmers, landowners, forestry bodies and county councils amongst others. We have also provided advice to the RSPB on the design of new plantations on its Lakenheath reserve.

The Golden Oriole Group has taken a number of initiatives to promote a wider appreciation of this special bird:

- a regular slide presentation to clubs and societies, which includes a video of birds at a nest
- articles to the Cambridgeshire, Norfolk and Suffolk Bird Reports
- participation in radio and television broadcasts by the BBC, Anglia TV and Channel 4
- creation of a website, www.goldenoriolegroup.org.

Paul Mason, the Chairman of the Golden Oriole Group, was presented with the RSPB President's Award in 1995, in recognition of his work on orioles in Britain.

LIST OF GOLDEN ORIOLE GROUP MEMBERS

GOG members past and present are listed below. They are referred to in the text by their initials as follows: AB Alastair Burn; AJP Tony Prater; AL Alan Livingstone; DH David Hopkins; DM Digby Milwright; JA Jake Allsop; JB Jeff Baker; JK Jeff Kew; JL John Lovett; KW Kevin Warrington; MJ Mike Jeanes; MR Malcolm Rains; PD Peter Dolton; PH Paul Holness; PM Paul Mason; NS Norman Sills; TM Terry Murfitt.

CHAPTER ONE

The Golden Oriole and the genus *Oriolus*

The male Eurasian Golden Oriole *Oriolus oriolus* is arguably one of the most strikingly colourful of all British breeding birds, vying for top position with the Kingfisher, Common Redstart, Jay and Eurasian Goldfinch, which makes it all the more frustrating that it is not an easy bird to see. It is well acknowledged that, for a brightly coloured bird, the yellow and black of the male Golden Oriole is most difficult to see when set against the high arboreal habitat it frequents, especially when the sun shines through the canopy foliage, creating a dappled effect. Usually song or call is the best aid to locating the bird. In fact, the species is not as retiring as imagined. However, its various stages of plumage change can be confusing at times. The Golden Oriole is slightly larger than a Eurasian Starling but smaller than a Mistle Thrush.

Full-adult males are a brilliant golden shade of yellow on the body, head, underparts and undertail-coverts. The lores are black, with the proviso that both eye-stripe and lores can vary in males. While most are dense black in front of the eye and extend only a small amount behind it, some variations have been noted. For instance, a photo by Sabine Baumann (*in litt.*) shows black only in front of the eye, which does not reach the rim of the eye, and there is no black at all behind it. As this was a ringed bird used in her telemetry work, it was known to be at least six years old. Baumann has also seen another male with a black pattern around the eye almost equal to that of the closely related African Golden Oriole (pers. comm.). This would be both in front, above and below, and extending much further behind the eye, as well as being a very strong black.

The wings and tail are a shiny black. On the tail the outer feathers have prominent yellow tips, which show well, and thinner tips on the central feathers, which usually only show when the tail is spread. The primary coverts, when spread, reveal broad dull-yellow tips, especially in older birds, and these form a distinct patch when the wing is closed. The bill can vary from a dark pinkish-red to near bright red. At nearly 28mm, the bill is longer relative to the overall average length of the bird (24cm), than that of a Eurasian Starling (20–21cm).

First-year males are predominantly green where older birds are yellow, and are dull black where the adults are shiny black. They mostly have few or no breast streaks, but there can be an intermediate stage in the second year when they are predominantly yellow but retain a greenish tinge, especially on the back. They need to be seen well to determine this. First-year males have been recorded as females at ringing stations on at least two occasions, once at Dungeness in Kent and again at Gibraltar Point in Lincolnshire. The Dungeness bird was subsequently trapped again at Dungeness the following year, by which time it carried the yellow plumage of a male. The Gibraltar Point bird was ringed and described as first-year female in May 1986 and found breeding in Cambridgeshire in 1996–97. At this stage it was positively identified as a male.

Second-year males are usually as yellow as a full-adult male but may still carry a green tinge on the back, scapulars and mantle. However, this is difficult to ascertain unless the bird is seen well. The wing-patch formed by the primary tips, which is yellow in adults, is still pale at this stage. It has taken quite a lot of nest photography, together with some pictures taken of birds caught at migration points, to appreciate this, and we suggest that this may be one of the ways to age some males. Underparts of these second-year males often have a paler yellow appearance than in older birds.

Females and juveniles are easily mistaken for each other. However, this initial confusion is understandable as the female is basically a bright green with a yellowish-tinged breast, especially towards each side under the wings, and greyish-white underparts. The wings and tail are a dull black and the rectrices have green-yellow tips where males are yellow. The lores are more of a diffused dark grey and the bill is a pale dull red. There are varying amounts of grey streaking on the breast, which gradually disappears with age, as much older birds, perhaps around eight to nine years, gradually take on a more male-like yellow plumage. These older birds still retain the other features of younger females. The legs and feet range from dark slate grey to a lighter grey. Juveniles are basically like females, but plumages change in stages and can be confusing until adulthood is achieved over a two-year period. After the fledgling stage, when it still retains some downy feathers, a juvenile bird has a cleaner green look until its first moult. The primaries and secondaries may be dull brown/black (and do not have a scaly appearance), but the secondaries are edged greenish and the wing-coverts, mantle, nape and crown are olive-green with pale feather edging, giving a scaly appearance. A ringed bird of known age (49 days) was recovered after flying into a window in Burstall, Suffolk. This preserved bird has a dull white throat, the breast and underbelly streaked grey-black, with yellowish flanks and vent. The back of the head, neck and wing-coverts are dull green, and the primary and secondary feathers are blackish, the primaries edged light grey. As at all ages, the rump is yellow. The undertail-coverts are yellow. The legs are a very pale grey and the bill appears black (but both have faded since death).

Adult and young females are indistinguishable and mostly have a bright green body, head and breast, the latter being brown-streaked quite prominently on young birds but gradually fading as the bird ages, leading into a creamy white under the belly. The undertail-coverts are a brighter yellowish-green. The wings are a dull black, as are the rectrices, which have similar markings as in the male but with duller yellow tips. The primary coverts are tipped pale yellow, forming a wing-patch, which is never yellow, unlike in young males. The lores are an indistinct dull grey. The bill is at all stages a dull (sometimes light) red, and the legs pale grey. As a female ages she slowly loses the breast stripes as the green body plumage gets more yellow, and finally becomes nearly as yellow as in the male. An old female can easily be mistaken for a male in flight. However, the black plumage remains dull and the lores are still a faint grey.

The nestling has rather short but plentiful down of a buff white to very light grey colour. It is also sometimes described as of a sandy hue. This gradually develops until at the fledgling stage, when feathers become fluffy and down is still hanging on, especially round the head and back. The eyes are dark but the colour of the lores is indistinct. There is a tinge of yellow over the eyes, under the wings and on the back.

In general, Golden Oriole plumages of both sexes can be confusing at all stages, since there are both green and yellow males and green and yellow females, according to age. However, close observation by the Golden Oriole Group and handling of birds of known age by ringers confirm that young males have white primary tips that turn deep yellow in older males, although the age of transition remains unclear. Old males and old yellow females may usually be separated by the colour of the lores – greyish in females and clear black in males. Nevertheless, some old females may show quite black lores; a good example is given by England (1964). It must be said, however, that many of the features described above have to be seen well to ascertain ages and sexes with any degree of accuracy.

CONFUSION SPECIES

The bird's flight style is noticeably undulating over open ground, sweeping upwards as it enters tree foliage, and in silhouette it can easily be confused with a Green Woodpecker or even a Great Spotted Woodpecker; the main difference is that a woodpecker will usually cling to the trunk of its target tree, something that orioles never do. Confusion with a Green Woodpecker is more likely when the bird is flying away from an observer, because this species has a yellow rump showing against dark wings and a green body.

Confusion with some thrushes and the Eurasian Starling is also understandable, as their wings too seem to be set forward, but the oriole's comparatively longer tail should remove any confusion. The same confusion is likely to occur with other green-type woodpeckers, such as Grey-headed Woodpecker in Continental Europe and Levaillant's Green Woodpecker in North Africa.

MOULT

The adults have a partial postnuptial moult before migration, beginning in late June and July in females at least. In addition to moulting body feathers, some birds then moult a few inner secondaries and some central rectrices. The adult birds then complete their moult from November to early February in their winter quarters. A few spring birds have two generations of feathers among their secondaries, although it is possible that the more worn secondaries are those that were renewed in the previous summer and are thus four to six months older than the feathers renewed in the winter. Worn secondaries are more usually present in second-year birds and occasionally in third-year birds. The time taken to complete the moult is known from captive birds to be about 70 days (Ginn & Melville 1983).

Young birds typically undergo a partial body moult shortly after fledging, but are unusual in that they have a complete, or almost complete, moult during the winter, although the plumage appears unchanged, making post-moult first-winter and second-winter birds practically indistinguishable. The moult strategy of orioles is unique among passerines with sexual dimorphism (Svensson 1992).

Adaptations to arboreal existence

The oriole is well-adapted to its arboreal habitat and is surprisingly acrobatic in its movements through the canopy. It has been recorded moving and feeding head-down (Wassmann 2004), a posture regularly adopted by other passerines such as tits, warblers and thrushes.

The need for a strong grip is met by the structure of the bird's foot, on which the outer and middle toes are lengthened to enable the bird to grip vertical stems without difficulty, a feature which Wassmann (2004) aptly describes as 'wie Steigeisen' – like 'climbing irons', i.e. crampons.

Longevity

The oldest known wild Golden Oriole is an 11-year-old male recovered in England. A record of a 15-year-old bird given by Rydzewski (1978) has since been found to have been an error (Akos Moniki, pers. comm.). This compares with males kept in captivity, which can live for up to 20 years (Zarth 1968).

Speed of flight

Wassmann (2004) gives records of flight speeds between 44.2 and 46.8km/hr (mean 45km/hr) in normal undulating flight, increasing when chasing a rival to 66.5–69.0km/hr, depending on wind conditions. He comments that higher speeds up to 80km/hr seem probable. There is also a report that an Indian observer estimated a flight speed of 40km/hr achieved by the closely related *kundoo* oriole flying alongside his car.

Biometrics

Little work on biometrics has been done by the Golden Oriole Group since we did not wish to disturb our breeding orioles unnecessarily. At the time we did not know how they would react to human interference. As time went on we found them to be quite tolerant of people, especially where sites were close by human habitation or where they must have seen agricultural operations going on in fields. However, because the nests were mostly very high up in tall poplar trees it was impractical for climbers to go up to nests more than once, that being to ring pulli.

Wing measurements (mm) in *O. oriolus* and *O. kundoo*

	Species	*Age*	*Mean*	*SD*	*n*	*Range*
Male	*oriolus*	Adult	153.9	2.9	52	148–160
	kundoo	Adult	145.1	3.1	17	139–150
	oriolus	Juv	148.3	4.8	11	141–158
	kundoo	Juv	144.5	5.8	43	139–167
Female	*oriolus*	Adult	151.2	2.6	27	146–156
	kundoo	Adult	142.8	6.9	6	139–155
	oriolus	Juv	150.1	3.2	8	144–155
	kundoo	Juv	141.9	4.2	37	136–152

Weights (g) of *O. oriolus* and *O. kundoo*

Period	*Sex/species*	*Mean*	*SD*	*n*	*Range*
Apr-May	M *oriolus*	67.8	9.62	8	51–85
	M *kundoo*	61	4.75	56	52–64
	F *oriolus*	69.2	2.17	5	67–72
	F *kundoo*	60.7	3.27	40	52–69
Aug-Sep	M *oriolus*	68	9.49	4	56–79
	M *kundoo*	74.1	6.35	4	66–80
	F *oriolus*	75	4	2	71–79
	F *kundoo*	70.3	8.54	12	55–80

Table 1. Comparison of wing lengths and weights in *Oriolus oriolus* and *O. kundoo*, showing size differences between the two species. Data for *O. oriolus* from Cramp & Simmons 1993; for *O. kundoo* from data collected at the Chokpak Ringing Station, Kazakhstan, in Gavrilov 2007.

THE GENUS *ORIOLUS*

The genus *Oriolus* comprises some 27 species, and there are three species of Figbird in the related genus *Sphecotheres*. *Oriolus* is an impressively widespread genus, with species found across the Palearctic region, from Western Europe to Siberia, in much of Africa, and in Asia from India through Southwest Asia to China and Japan. More species are scattered throughout the Philippines, Indonesia and into northern Australia and down the east coast. They are collectively known as the Old World Orioles. Close to 70 subspecies, many of them island endemics, are recognised, so it is not surprising that the taxonomy of the Oriolidae is somewhat fluid. In the last two decades, a number of subspecies have been given full species status, notably the Indian Golden Oriole, *O. kundoo*, formerly a race of Eurasian Golden Oriole. The wider relationships of the genus *Oriolus* are still not resolved. Traditionally considered to be related to crows and drongos, recent work on DNA puts them close to the cuckoo-shrikes.

The reason such taxonomic issues are as yet unresolved is that a great deal of the science remains to be done. Until molecular genetic analysis is carried out, much of what we perceive about the relationships within and outside the family will continue to rely on conventional characters such as osteology, plumage and voice. In fact, there is what one might call 'a conflict of interest' between the layperson and the scientist in this respect. The layperson is happy with these simple groupings: the yellow-headed group (*O. oriolus*, *O. kundoo*, *O. auratus*, *O. chinensis* and *O. tenuirostris*); the African black-headed group, (*O. chlorocephalus*, *O. crassirostris*, *O. brachyrhynchus*, *O. monacha*, *O. percivali*, *O. larvatus* and *O. nigripennis*); the Asian black-headed group, (e.g. *O. hosii*, *O. cruentus*, *O. traillii*, *O. mellianus* and *O. xanthornus*); and the 'Mimeta' or Brown Oriole group (*O. szalayi*, *O. phaeochromus*, *O. forsteni*, *O. bouroensis*, *O. melanotis*, *O. sagittatus*, *O. flavocinctus*, *O. xanthonotus*, *O. steerii* and *O. isabellae*). For the scientist, on the other hand, these groupings depend on superficial characteristics such as plumage and voice, *i.e.* characteristics that enable identification of the birds in the field, but which may not reflect at all the true evolutionary relationships within the oriole family tree. Only molecular genetic analysis can resolve these questions.

The Eurasian Golden Oriole *O. oriolus* is the most widespread of the entire *Oriolus* genus, probably the most numerous, and certainly the longest-distance migrant. The Indian Golden Oriole *O. kundoo* which we are considering as a species separate from *O. oriolus*, within which it was formerly placed, breeds as far north as southern Kazakhstan into Kyrgyzstan, and winters in southern India and Sri Lanka. The migration distance is not as far as that of *O. oriolus* but it probably flies to the highest altitude on migration when travelling north–south over the Himalayas. Figure 1 indicates the distribution of *oriolus* and *kundoo*.

The second most widespread is the Black-naped Oriole *O. chinensis*, which is found throughout eastern Russia and to Southeast Asia as far as China and Korea, and to Japan on migration. There appear to be three variants of this species: one is sedentary in the south of the range; one migrates north to breed; and a third is a short-distance migrant from south China, Laos, Vietnam and north Thailand, wintering in Malaysia, Singapore and Sumatra. The Slender-billed Oriole *O. tenuirostris* was until recently regarded as a race of Black-naped Oriole.

Of the remaining species, the Asian Black-headed or Black-hooded Oriole *O. xanthornus* and the African Black-headed Oriole *O. larvatus* vie for third place. In fact, in Africa there are six more species: African Golden Oriole *O. auratus*, Black-winged Oriole *O. nigripennis*, Montane or Brown-tailed Oriole *O. percivali*, Dark-headed or Forest Oriole *O. monacha*, Green-headed Oriole *O. chlorocephalus* and the endemic São Tomé or Great-billed Oriole *O. crassirostris*.

Many of the species in Southeast Asia, Indonesia and the Philippines are little studied, with the exception of some endangered species such as the Isabela Oriole *O. isabellae* and Silver Oriole *O. mellianus*, or locally vulnerable species such as the Dark-throated Oriole *O. xanthonotus* and the Black Oriole *O. hosii*, the latter

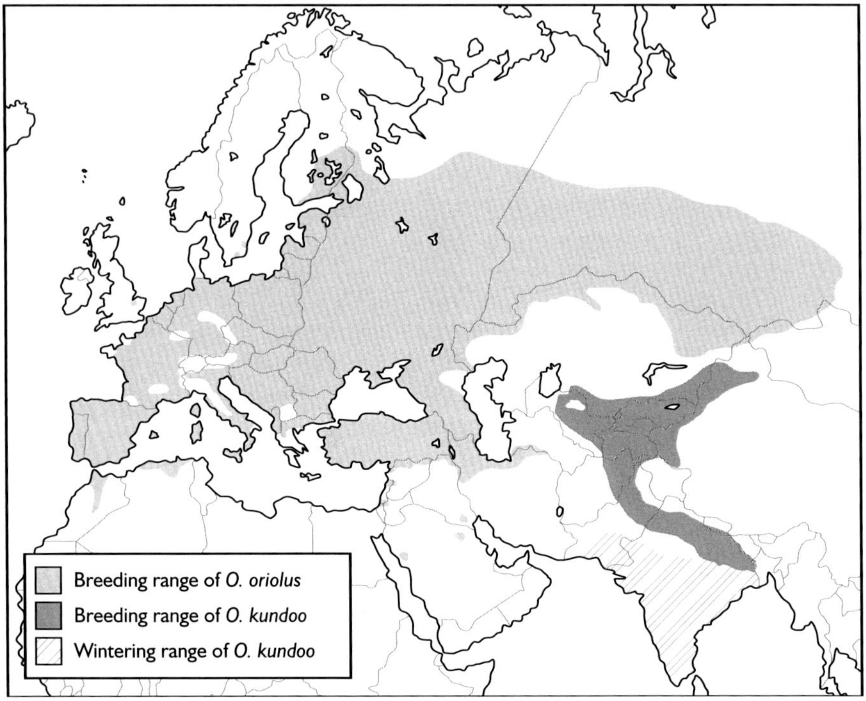

Figure 1. The breeding range of *Oriolus oriolus*, and the breeding and wintering ranges of *O. kundoo*.

being endemic to Borneo. Australia has two oriole species: the Olive-backed Oriole *O. sagittatus* and the Yellow Oriole *O. flavocinctus*. The closely related figbirds occur in north and north-east Australia and associated islands, as well as in southern New Guinea. Three species and a handful of subspecies are currently recognised; the Green or Timor Figbird *Sphecotheres viridis*, the Wetar Figbird *Sphecotheres hypoleucus*, and the Australasian Green (or Yellow) Figbird *Sphecotheres vieilloti*.

A list of the *Oriolus* and *Sphecotheres* genera, based on the most recently published checklists (Sibley & Monroe 1993; Dickinson 2003a), is given in Appendix 1.

The yellow-headed group

In addition to the Eurasian Golden Oriole *O. oriolus*, there are four other species in this group: Indian Golden Oriole *O. kundoo*, African Golden Oriole *O. auratus*, Slender-billed Oriole (*O. tenuirostris*) and Black-naped Oriole *O. chinensis*.

The Indian Golden Oriole, *O. kundoo*

The Indian Golden Oriole was until recently regarded as a race of *O. oriolus*. Rasmussen & Anderton (2005) treated *kundoo* as a separate species in *Birds of South Asia*; they say it does not seem to interbreed with *O. oriolus*, even though they breed in close proximity towards the north of *kundoo*'s range. They also believe that differences in morphology and vocalisations are 'comparable or greater than those between other oriole species'. *Handbook of Birds of the World* (in prep.) will feature the split.

Identification

The plumage of *O. kundoo* is very like that of *O. oriolus*, but the bird is smaller and its bill longer in relation to its size. The eye-stripe is black but extends further behind the eye than in *O. oriolus*. The carpal patch on the folded wing is larger, and the yellow on secondary and tertial tips covers a slightly larger area. There is a greater proportion of yellow on the tail than in *O. oriolus*, especially when seen from below. The outer rectrices are also yellower. The bill in usually a paler red than in *O. oriolus* and is more curved. This general description is supported by a report of birds seen during a birding trip to Kazakhstan: 'The birds we saw corresponded to *kundoo*, in which the black loral line extends a short way behind the eye and the intensity of the yellow pigment in the plumage is slightly richer than that of the nominate form. Females also differ in being yellower both above and below and exhibit reduced streaking on the underparts.'

The female is very similar to the female *O. oriolus* but the feather streaks below are more obviously sharper. The black-billed juvenile is more heavily streaked on the belly and has a bright yellow face and upperparts, whereas juvenile *O. oriolus* is olive-grey. The latter is also lacking a prominent yellow supercilium, which is obvious, along with a long, dark eye-patch, in *O. kundoo*. There is one possibility of some confusion between juvenile *O. kundoo* and juvenile Slender-billed Oriole where both occur, but *O. kundoo* juveniles are more heavily streaked black on the breast and underbelly. The beaks of the juveniles are black. Biometric data given in Rasmussen & Anderton (2005) is as follows: length 183–203mm, head 54–56mm, tail 87–97mm.

The Indian Golden Oriole winters in southern India and sometimes in Sri Lanka. It migrates north to breed in northern India, north Pakistan, Nepal, Afghanistan (mainly recorded around the capital, Kabul, but also near to other large towns), Turkestan, the lower mountain ranges of Tajikistan, Uzbekistan (where they seem to be quite plentiful, even in some desertified oases) and southern Kazhakstan. Some have been noted in the western Tien Shan mountain slopes of Kirghizstan and into Xinjiang in north-western China; there they breed on the southern slopes of the Tien Shan mountains, while Eurasian Golden Orioles breed on the northern ones (Ma Ming, *in litt.*). Birds are seen in Tibet but breeding there is uncertain. In northern India (where Indian Golden Oriole is claimed to be resident all year) and Pakistan, it is recorded up to 3,000m on migration, but the numbers thin out north of the Himalayas, where it is not found above 1,800m. However, the main breeding range is in both Indian and Pakistani Kashmir through to western Bangladesh. Its arrival back on breeding territories coincides with the ripening of mangoes in April in northern India, giving rise to its local name, mango bird, because it also blends in with the colour of the fruit and the green and yellow mosaic pattern of mango tree leaves. In India it is called *peeluk*, 'the yellow one'.

Pakistan

Kashmir represents the main Pakistani breeding range, although a few do breed in north Baluchistan province, where Meinertzhagen (1920) claims to have found some breeding in orchards near the town of Quetta. Others have found some nesting along the Zhob river valley and the Shingar mountains. In Northwest Frontier Province they become more common. Many trip reports comment that the birds are encountered in the Islamabad area. In the north Pakistani valley of the river Yarkhim above Chitral, breeding orioles arrive in late March and early April. The males set up and announce their breeding territories by singing from high walnut trees along otherwise conifer-lined lower mountain sides. David Debb (pers. comm.) described how at a village where he spent time teaching, the schoolchildren looked forward to the return of the orioles as a pointer to spring and a good year ahead.

The Kurram valley also has birds nesting in its upper reaches, as do the Baltistan, Gilgit, Panjorka, Swat (2,400m) and part of the Kaghan valleys. Indian Golden Oriole was also found to be breeding at 2,600m in the Babusar, Chilas and Gilgut areas. South of these places breeding thins out considerably. In Pakistan and India, the birds prefer well-wooded and thinly forested areas that are moderately open and dry. Nests are usually placed very high in trees such as Walnut, but some have been recorded in loquat (at 1.5m in one case) and in silk cotton. However, at Rohtak, west of Delhi, a Mr F.R. Blewitt (quoted in Hume 1889) claimed he had taken more eggs from nests in Tamarind than any other tree. At Lucknow in Uttar Pradesh Indian Golden Orioles have been recorded in a Neem tree, sometimes as low as 3.1m. Another nest-taker climbed high in a banyan. He also took two more nests, one of which was in an acacia, the other in Loquat, these again at low heights. Yet another record at Allahabad was in a mango. Whether it happens by coincidence or not, nests have been recorded in trees also occupied by the Black Drongo. It was thought by several observers that they might gain some protection from this species' aggressive actions towards predators and others invading its territory.

India

Whilst *O. kundoo* can at one time or another be found almost anywhere in India according to season, its breeding range is basically in the northern provinces, from Indian Kashmir to western Bangladesh and, as in Pakistan, venturing into highland valleys, sometimes to 2,500m. A very small number breed in the region of Mumbai (formerly Bombay), and numbers are augmented during winter from around early October. There is local movement in winter, which is thought to be within the region. There is also a further influx of this species during winter to Mulund, Colaba, Sion or Kandivli and the National Park, but a pair or two is always possible and the species could be breeding somewhat more widely. In fact, until the 1980s Indian Golden Oriole had been considered just a winter migrant to the region until a pair was discovered breeding at Kandivli:

> Interestingly also, I have noticed that when the Orioles have their fledglings up and about and moving with the adults, [the adults] are particularly effervescent, and it is quite possible this might have been so at Mulund too. There might have been a nest somewhere close-by and the young are learning the tricks of the trade and the trials of life around Aug–Sept.
>
> (Abdulali 1981)

The birds winter throughout India, from Gujarat to the Bay of Bengal and in Sri Lanka. However, the Great Indian Desert near to the Pakistan border seems to be avoided. The orioles are is found generally in the plains of India and lower ranges of the Himalayas, up to 1,400m. This is a good deal higher than its congener in Europe. Elsewhere, the oriole is a rare breeder in western Bhutan. There are also reports from the Maldives and Andaman Islands, far out in the Pacific Ocean and Bay of Bengal respectively – all of females, though perhaps they were immatures as these are possibly more likely to overshoot.

Kazakhstan

Indian Golden Orioles reach to the south-east corner of Kazakhstan and are common around the main city in the country, Almaty (formerly Alma-ata).

Parks and gardens are used extensively by orioles in Kazakhstan, as are almond and walnut orchards, such as in the Panfilov Guards Park, where one OSME trip report in May 2004 met with a pair on the 10th:

> At least four at the arboretum to the east of Almaty on 12th, a pair in the same area again on 14th, a male at the farm near the Konshengol camp on 15th and one heard in the orchard near Astana on 18th. The birds we saw corresponded to *kundoo*, in which the black loral line extends a short way behind the eye and the intensity of the yellow pigment in the plumage is slightly richer than

> that of [*O.oriolus*]. Females also differ in being yellower both above and below and exhibit reduced streaking on the underparts.

Indian Golden Oriole has been recorded breeding in the lower reaches of the Syr Darya. However, the larger part of the country lies within the range of *O. oriolus*. Both species are regularly trapped at the Chokpak ringing station run by Edward Gavrilov and his teams. He reports that to date more *O. kundoo* are ringed in spring than autumn (with *O. oriolus* it is the other way round), which would indicate a migratory pattern of northward travel on a broad front with a more restricted route southwards, again possibly following a fruit supply along the lower farmed reaches of the mountains back to the birds' winter quarters.

Afghanistan

Afghanistan is thought of these days as a war-torn, arid, mountainous country denuded of trees, and in part this is true. However, there are forested areas and also lines of trees along irrigation channels, and many of these consist of poplar and willow. The native poplar is extensively used in constructing the flat-roofed dwellings inhabited by most Afghans. It is laid as planks on rafters and covered with other plant material and soil to provide insulation from the elements. Fruit orchard crops reportedly cover 10% of the irrigated area, which in itself is not a very great area of the whole country. Grapes, almond, apricot, pomegranate and apple occupy some 87% of the area devoted to fruits. Mulberry is a primary fruit crop, with this normally grown along roads and boundaries and Walnut is also an important crop. All these trees are well-known to provide nesting sites for Indian Golden Orioles. Kullberg (2002) visited the country in 2002, but because of the need for personal security all observations had to be made from within travelling vehicles. However, it was noted that orioles were common in Kabul, the capital, in the east and around Herat in the north-west. Other reports mention Jelalabad on the Kumar River section in Afghanistan. On his journey through Pakistan to get to Afghanistan, Kullberg (2002) reported seeing the birds in good numbers around Islamabad.

China

The ranges of the two species, *O. oriolus* and *O. kundoo*, are adjacent to one another but do not seem to overlap in Xingjiang province in northwest China. *O. kundoo* breeds in oases in the south and *O. oriolus* in the north, with the dividing line being approximately along the Tien Shan mountain range to the state capital of Urumqi. These oases are basically small plantations and belts in agricultural land. The birds mostly nest in poplars, *Populus nigra* and local hybrids suitable to the local conditions, which are often in commercial plantations and shelter belts, and *Populus alba*, which grows wild in the region.

Indian Golder Orioles have been found to an elevation of 2,130m in the Kunlun Shan Mountains, and probably occur to 3,000m. These are in the south of the province bordering the northern ranges of the Himalayas. The birds are regarded as common (Ma Ming, pers. comm.).

Indian Golden Orioles breed in the gardens of Karakul, Bukhara, Kettakurgan and Samarkand oases, in mountain sayas of Nuratau, Zerafshan and Turkestan mountain ridges where there are deciduous tree plantations. In mountains, breeding reaches to a height of about 2,200m in the vicinity of Iskanderkul Lake. Near Samarkand, Sagitov (1965) recorded the spring arrival of male orioles in the period 1959–63, earliest date 29 April, latest 7 May. In a particular study carried out by Sagitov in 1958, 11 pairs, which bred in Amankutanskaya forest block, were involved in breeding activity by 30 May. Nests were built in several different tall trees, and those particularly noted were poplar, apple, walnut, willow, apricot, and an *Ailanthus* spp. tree. Here nest construction was thought to be carried out by both sexes. And if so, this will be different behaviour from that of *O. oriolus*. In some cases nest-building went on for quite a prolonged period. One pair in the Amankutan forest had begun to build on 17 May but did not finish till 27 May. On 29 May, the first egg was laid, and the young fledged on 19 June.

There appears to be a difference in the timing of egg-laying between the Zerafshan valley in the lower part of the country, where it occurs around 10–20 May, and the mountainous regions, where 1–10 June is the norm. During the first days after fledging, the young birds stay near to the nest while their parents continue to feed them. After about four days, the young birds leave the nesting territory together with their parents, and this is when family parties are often to be seen in gardens, pecking out cherries and grapes. Oriole insect food here is quite diverse. It consists of larvae of Cerambycidae, Cephidae, caterpillars of the Black-veined White Butterfly, Staphylinidae, Curculionidae, Melolonthidae and other butterflies. Undoubtedly, *kundoo* is an extremely useful pest control for the horticultural and forestry industries (Gavrilov, 1974).

In Kirgyzstan much use is made of orchards and groves in the lower regions, where they are common. Similar conditions exist in Tajikistan along river valleys from the lower reaches to mid-way elevations. One bird ringed in India was recovered during the breeding season in the Vakhsh river system 25km south-east of the capital, Dushanbe. It was nine years old at the time of its death and must have been regularly breeding at this place.

Nepal and Tibet

In Nepal, breeding has been confirmed along the Kathmandu valley and near Dhankthuta (Inskipp & Inskipp 1985). The bird is found in trees and groves around cultivated areas and other wooded areas to an elevation of 1,830m in some places. Many of the birds seen are on passage to Tibet and western China in spring and returning in autumn. Other *Oriolus* species found here are the

Black-hooded, the Maroon and, more rarely, the Slender-billed. In Tibet, it is only a passage bird through the Outer Plateau of Tibet *en route* to Xingjiang in spring returning in autumn with late dates of 20–21 September (Vaurie 1972) Records from Burma, where it could logically winter, are hard to come by, but here and in south India and Sri Lanka it would meet up with the Asian Black-headed Oriole.

Where the exact line lies between the limits of breeding *O. oriolus* and *O. kundoo* is a little unclear, but it would be reasonably safe to assume that the western edge of Indian Golden Oriole's range would be close to the eastern shores of the Aral Sea. According to Dementiev & Gladkov (1954), *O. oriolus* breeds north of a line from the Mangyshlak peninsula on the eastern shore of the Caspian Sea, south of Astrakhan, to Karagaly near Qharagandy, north of Lake Balqash. It then dips south-east to the eastern Altai mountain range close to Urumqi in the Xingjiang province of China and just into the western edge of Mongolia.

The other three species in this group are closely related to the Eurasian and Indian Golden Orioles.

The African Golden Oriole, *O. auratus*

The African Golden Oriole *O. auratus* breeds from the West African coast to the east coast, south to Mozambique. In essence, its range lies right through the centre of Africa south of the Sahara, but in the winter months of August to April it is known to more across a wider belt, but not to the far south of Africa. It can be confused with *O. oriolus* if not seen well, especially in the case of females. Habitat preferences are miombo woodland but African Golden Oriole will use the closed canopy forest around rocky hills and riverine gallery forests. It is less often seen in gardens and dry savanna (Feather 1986). Food is mainly as other orioles but it has been seen to take locusts (Roberts 1928). Brooks *et al.* (1987) record the bird catching flying termites and returning to a perch to consume them. As to fruit, Oatley (1969) saw it feed on unripe fruit of *Euchlea* spp. and also those of the non-indigenous Neem tree. The nest is similar in construction and materials as those of other African orioles. Baumann (2000a), however, records that it is usually placed in the central branches of a tree while the nest of Black-headed Oriole is mostly at a lower level.

The Slender-billed Oriole, *O. tenuirostris*

The Slender-billed Oriole *O. tenuirostris* was formerly considered a race of Black-naped. It too has a black nape, but the eye-stripe is much thinner than that of Black-naped, and it also has a distinctly narrower and longer bill. It is found from eastern India through south Bhutan, south-west China and Burma to northern Laos and Vietnam. In Bhutan its range is restricted to the eastern valleys between

1,200m and 1,600m. Between November and April it migrates a short distance south and west and can also be met with in Nepal in winter. That it is closely related to Black-headed is illustrated by incidents of reported hybridisation.

The Black-naped Oriole, *O. chinensis*

The Black-naped Oriole *O. chinensis* is found over a wide area, from Vietnam to Malaysia to Indonesia to the Philippines and the Nicobar/Andaman island chain. Birders visiting the Beidahe region of China in May report large numbers moving north. Some are believed to breed in nearby poplar plantations (MR, pers. comm.). The first birds reach the Iman river by about 14 May, and in significant numbers by four days later, when trees are already in bloom. In Beijing, the birds' arrival is probably timed to coincide with the blooming of cherry trees. In the plains, the birds breed in oak woods, while around towns and villages, they are to be found in poplars.

At 26–27cm long, Blacked-naped oriole is one of the largest orioles. The plumage of the male includes a bright yellow or orange-yellow body, which includes head, back, neck, underparts and uppertail. Flight feathers are black edged with yellow; the main tail feathers are also black. There is a small yellow wing-patch. The most prominent identification feature is a wide black eye-stripe extending around the back of the head. In some subspecies this black band can give the appearance of a bandit's traditional mask. Females and juveniles have roughly the same plumage details except that the body has a greener tinge. MacKinnon & Phillips (1993) note the flight pattern to be undulating, with a conspicuously slow but powerful wingbeat. To date it is probably the most intensively studied Asian oriole. According to the most recent taxonomic thinking, there are 20 subspecies. What appears to be the most widespread race, *diffusus*, winters in areas of eastern India, Burma and Bangladesh. It migrates north to China and Korea, and to the south-east coastal region of Siberian Russia. It also winters in very small numbers on Honshu, Japan.

In Russia, Dementiev & Gladkov (1954) state that breeding occurs along the Argun river, and the bird is fairly common in the valley of the Amur river between the towns of Blagveschenk and Khabarovsk. It is reported to be very common in the Ussuri river basin, but its range from the Sikhote-Alin mountain range down to Vladivostok is unclear. In south-east Manchuria, it is said to be very common, even in the high Da Hinggan Ling Mountains in the west. It also breeds in good numbers on the Korean peninsula.

In the second half of May, the birds arrive in the Vladivostok area, when the traditional courting behaviour of all orioles results in pair formation by early June. Black-naped Oriole hammock nests are possibly the largest of all the *Oriolus* species. They are often recorded at only 2–5m from the ground in young trees, but even in older trees they are placed in drooping branches at no higher than 5m. Trees used are mainly poplar, oak, willow and bird cherry; usually the younger trees are chosen. Nests are certainly larger than those of Eurasian Golden Oriole but otherwise show little difference in shape. The clutch size is three to four. In Singapore the nests are built high in trees at the end of branches. Two to three bluish-white eggs speckled brown are laid. In addition to the usual oriole diet of fruit (especially figs) and berries, the birds take small animals and insects and occasionally the nestlings of other birds. There are at least two records of *O. chinensis* further west towards the Baikal region; in 1975 (Dorzhiev 1996) and in 1985 (Melnikov 1998), which may indicate that it is more widespread in eastern Siberia than previously thought. It is possible that within the limits of the Irkutsk-Cheremkhovskaya Plain there is a contact zone of *O. chinensis* and *O. oriolus*, though both species are extremely rare here. It seems that at the present time, however, there is a gap between the ranges of these species that does not exceed 80km (Melnikov 1998).

Birds arrive back in their breeding areas in the third week of May, usually announcing their presence by means of the typical oriole 'catcall' (sometimes referred to as 'chaying'). The earliest record was from 20 May, the latest one 27 May, the average being 25 May. Departure is characterised by a decline in the intensity of calls and by complete cessation of song. The latest record, registered by song, was 23 August in 1986. The orioles begin to nest at the end of May and early June. Nests are placed in tall trees, usually white birch, in the top third of the tree, where they are sometimes still to be seen in the following spring, this enabling fieldworkers to conduct spring counts of nests from the previous year. Nests of separate pairs can sometimes be as little as a few tens of metres apart. These 'Chinese' Orioles, as they are known to Russians, begin to undertake nomadic movements in July, often in family parties of between three and eight in number.

The males have a song like all orioles, which is a variation on a theme and is described as 'gri-gorii', and the female sometimes utters a 't-yu' call in repeated fashion. In their winter quarters they frequent open forests and plantations around villages and beach and mangrove swamps, keeping mainly to trees but occasionally dropping to the ground to pick up insects.

Being a widespread species that occurs on many islands of the Indonesian archipelago, the Philippine islands and the Malay peninsula, *O. chinensis* is quite complex taxonomically. To complicate the picture even more, only some of the races are migratory, and these can be either long-distance or, in India, short-distance. It is also said that birds breeding in Vietnam and Laos move to Singapore in winter but they are impossible to separate from resident birds on plumage characteristics. The nominate race was first described by Meinertzhagen in 1923; these birds occur on the Philippine island of Luzon, especially around the capital, Manila. However, there appear to be significant plumage differences between those in the south of the same island at Sorgoson and those on the island of Fuga off the northern tip. This gave three races for Luzon, all distinct enough to make them identifiable in the field (Gilliard 1950). There are two races on the Andaman and Nicobar islands, one for each part of the chain. In the Andamans, male birds are seen to have very narrow black nape bands and slender bills. As they are nearer to Slender-billed territory, they may be taxonomically closer to that species than to *O. chinensis*. In Sulawesi some separation occurs between birds in the north and those in the south. There the males have a narrowish black band extending from the lower nape through the eye and coming to a point behind a deep pink bill. Otherwise the plumage is bright yellow with black primary feathers, and the wing-coverts are light olive-yellow edged black. The tail is yellow with some central feathers black.

The other subspecies within the Wallacea region include:

- *O. c. sanghirensis*, found on Sanghihe, 25km north of the northern tip of Sulawesi, which sports a black hindcrown.
- *O. c. frontalis*, found on the Banggai Islands and Sula off the eastern mainland of Sulawesi, and *O. c. formosus*, found on small islands off the north coast of Sulawesi, which are similar in that the greenish-backed male has a black head, with the amount of yellow reduced to a little patch at the front of the head.
- *O. c. melanisticus*, found on the northernmost islands of Wallacea, the Tallaudes, which is similar to *formosus* but the male has a darker green back.
- *O. c. broderipii,* found on the most westerly of the Lesser Sundas, which has deep orange plumage and black wings and head markings.
- *O. c. boneratensis*, found on a group of islands to the south of Sulawesi but north of Flores, which is very like *broderipii*.

The birds in this region breed from February to June, the usual hammock nest generally placed in the mid-to-top section of any given tree. Birds utter loud far-carrying songs and calls, often sustained for long periods of time and in company with others of the species (Kennedy *et al.* 2000; Dickinson 2003b, 2004).

Those birds designated *O. c. diffusus* migrate north between late May and early July, and breed at a different time to those that stay on home ground, so to speak. Those in Java and Sumatra, *O. c. maculatus* for example, breed from February to June, and nests have been also recorded in August and December (MacKinnon & Phillips 1993), so the question arises whether they are double-brooded or if they merely have a propensity to breed nearly all year round. Scott Leatham (*in litt.*) wrote:

> *O. chinensis* was the bird that delighted me all the time I was in the East and it seems to be the species which has most successfully adapted from its natural habitat of lowland woodlands and mangroves to parks, gardens and cities. When I was last in Singapore, its mellow notes were a delightful counter to the roar of traffic, and its flashes of yellow ever present surprises in the suburban gardens of this ultra modern city state.

AFRICAN BLACK-HEADED GROUP

The Green-headed Oriole, *O. chlorocephalus*

The Green-headed Oriole *O. chlorocephalus* belongs taxonomically in this group, despite the fact that it does not have a black head. The three races of Green-headed are found in south-east Kenya, in the eastern half of Tanzania, in south Malawi and central and southern Mozambique (where the reserve at Gorongoza is reputed to be the best place to observe these scarce birds). Its preferred habitat is montane and submontane evergreen forest, ranging from 750 to 2,000m. The bird has not been greatly studied but is known to take fruit in company with other orioles of the Black-headed group, and there is a record of a bird taking flowers on a non-indigenous plant of the *Grevillea* genus (Clancey 1970). Few nests have been seen but one described by Maclean (1993) was 10m above ground and looked similar to the nests of other Black-headeds. Another was found 40m up in an African mahogany tree at a forest edge (Benson & Benson 1947). There is no official estimate of population size, but the population is believed to be large, given that the species is described as 'frequent' in some parts of its range. At the moment, it is not regarded as approaching the threshold for the population decline criterion of the IUCN Red List, i.e. a decline of more than 30% in 10 years or three generations.

The São Tomé Oriole, *O. crassirostris*

The São Tomé or Great-billed Oriole *O. crassirostris* lives only on the island of São Tomé, situated off Gabon. It has hardly been studied at all but is regarded as 'Vulnerable' because only a small population lives in a small area of suitable primary and mature secondary habitat. According to Birdlife International (Birdlife Species Factsheet, 2004), an improved knowledge of its requirements and population could release it from that status. At the moment the population estimate is between 250 and 999 birds in an area of 480km^2, with an estimated breeding density of 25ha per pair (excluding the unsuitable north-east of the island). The report goes on to say that large areas of its habitat were cleared for coffee and cocoa plantations, and now numbers of small farms are increasing, along with road improvements, which could lead to problems for the bird. This species is not found on the nearby island of Principe.

The Western Black-headed Oriole, *O. brachyrhynchus*

The Western Black-headed Oriole *O. brachyrhynchus* is a native of the rainforests of Guinea, Sierra Leone, Liberia, Ivory Coast, Ghana, Togo and Benin. The subspecies *O. b. laetior* is apparently restricted to the western and central 'Lower Guinea' rainforests, as found in Nigeria to East Kenya, Gabon, northern Angola and Congo.

The Ethiopian Oriole, *O. monacha*

The Ethiopian, Dark-headed or Forest Oriole *O. monacha* consists of two races, both found in Eritrea and Ethiopia. The Mountain Oriole *O. percivali*, sometimes referred to as Black-tailed, Percival's or Montane Oriole, is found in the dense montane forests of eastern Zaire, Uganda, Kenya and northwest Tanzania. Most birders have seen the bird during visits to the Bwindi Impenetrable National Park to watch the gorillas. On the western edge of the Uganda depression, this species has been known to reach 1,600m in the Mpanga Forest.

The African Black-headed Oriole, *O. larvatus*

The African Black-headed Oriole *O. larvatus* is widespread in East Africa, its five subspecies ranging from southern Sudan, Ethiopia and the coastal forests of Somalia through Uganda, Kenya and Tanzania to the Lake Victoria basin and eastern Congo. It is also found further south in Zimbabwe, Mozambique and eastern South Africa, and is also present in Angola and northern Namibia. It inhabits acacia, mangroves and other types of woodland. Where it is in competition with the Western Black-headed, it retreats to woodland edges and behaves similarly to the Montane Oriole. It builds a typical hammock nest, made with woven plants, spiders' webs, lichens and mosses, typical of all the Black-headed family. The nest is usually placed well away from the trunk of a tree and is often hidden in growing *Usnea*. These orioles have been known to use the same tree year after year. Vernon (1984) reports that the nesting success is 47%, with an average of 1.1 young fledging per nesting attempt in miombo woodland in Zimbabwe. The species feeds, probably like most orioles in Africa, on invertebrates such as dragonflies, honeybees, moth larvae, and these birds also take termites in the air (Baumann 2000b). This oriole also eats ripe and unripe fruit, including figs, as well as olives and the green, hard berries of *Randia rudis*. It has been recorded as taking nectar from the flowers of broad-leaved erythrina and from silky oak and aloes (Skead 1995). Hard, dry seeds of the Australian flame tree are also eaten (Skead 1995). Its predators include Peregrine Falcons and the Black Sparrowhawk (Tarboton *et al.* 1987), although Van Someren (1956) found a nest in a high tree which he thought to have been predated by a Tropical Boubou.

The Black-winged Oriole, *O. nigripennis*

The Black-winged Oriole *O. nigripennis* is found in the lowland forests of western and central Africa. It has hardly been studied in detail but is assumed to follow the habitat and habits of most of the other African orioles.

ASIAN BLACK-HEADED GROUP

The Asian Black-headed Oriole, *O. xanthornus*

The Asian Black-headed Oriole *O. xanthornus*, also known as the Black-hooded Oriole, is common in forests, small woods and even in townships in India and in countries around, and islands in, the Indian Ocean. It can also be found in northern Sumatra, coastal areas of north-east Borneo, and on some nearby islands. Not found in Vietnam, it is present westwards to Thailand. Regarding its biology in India, Hume (1889) writes that its breeding season lasts from April to the last days of August. The nest is the usual cup-shaped nest of the genus made from various fibres, which include grass, tamarisk and bark, decorated externally with 'odds and ends, bits of lichens' and other things. The nest is somewhat smaller than that of *O. kundoo* and is invariably placed in mango, mowah or bamboo trees at the extreme ends of an upper horizontal branch, which is overhung by many leaves. Both sexes build the nest. Within his text, Hume quotes one J. Davidson as saying that he found two nests in the same tree at the same time. It seems there can also be some pairs on eggs while others have quite old fledglings. Usually three eggs are laid, which, compared with the glossy whiteness of the eggs of *O. kundoo*, are less glossy on a pinkish background.

The Black Oriole, *O. hosii*

The Black Oriole *O. hosii* is considered by Birdlife International to be threatened. It is confined to Sarawak in Malaysia and is found only in mountain forests between 900 and 2,000m. It prefers moss transitional forests on the north side of the Dulit range, where the soil is sandy. MacKinnon (1988) describes the Black Oriole as 'a quiet yet active species, keeping to tree canopies and mostly found in small parties and located often only by clear downward-inflected whistles'. It is an all-black bird, apart from a chestnut-brown patch around its vent. Unfortunately, the areas used by this species are vulnerable to habitat loss and encroachment as a result of forest clearance. It is not found in the area of Mount Kinalabu, which is the most threatened by felling and palm planting, but this does not lessen fears for it and for the 27 other montane Borneo endemics (Brooks *et al.* 1999).

The Black and Crimson Oriole, *O. cruentus*

The Black and Crimson Oriole *O. cruentus* replaces the Black Oriole where the latter is absent in Sarawak, mostly above 1,000m. Again inhabiting montane and mossy forests, it confines itself to the canopy, making it difficult to locate. Its call differs from the Black Oriole, being more of a harsh, cat-like meow, 'Teee-ui' (Scott Leathard, *in litt.*) or 'hee-hee-hee-hiew' (MacKinnon, 1988). This bird also inhabits Sumatra, where it is said to be 'occasional', and Java and the Malay peninsula, where it is regarded as common. It travels mostly in pairs or singly, feeding on the usual *Oriolus* diet of insects, including termites, and fruit such as figs and other berries, but M Strange asserts that it joins 'bird waves' or mixed parties of several species flitting through the middle canopy of tall trees, stirring up insects so as to feed on them. The overall plumage is black but the male has a crimson-red chest and a similarly coloured wing-patch. In the female the chest is grey and there is no wing-patch.

The Maroon Oriole, *O. trailli*

The Maroon Oriole *O. traill*i is said to be quite a common resident from Burma to the Himalayan foothills up to 4,000m in Bhutan, Nepal and northern Bangladesh. It also occurs in Thailand, Vietnam, south-east China and the islands of Taiwan and Hainan. It is a striking bird of bright maroon and black plumage, and can often be seen taking nectar from flowers. The birds found on Taiwan are regarded as subspecifically distuct, *O. t. formosus*. When the bird is seen, it is quite obvious that the 'maroon' has a much redder hue compared with the nominate race. Here it is regarded as threatened because of the destruction of habitat going on all over the island, including the creation of several reservoirs.

Another race, *O. t. nigellicauda*, which is a visitor to Vietnam from its breeding grounds in south-west China, has a brighter maroon-red plumage and a red wash on the mantle, as well as a noticeably deep red rump and tail. *O. t. robinsoni*, found in southern Laos, shows only slight differences from nigellicauda.

In Bhutan and Nepal, Maroon Oriole is regarded as 'notably common'. It behaves in a typical *Oriolus* fashion, remaining in high deciduous forest canopies, often in association with other species. Like all orioles, it builds a hammock nest slung between branches of trees. Hume (1889) quotes descriptions by various Victorian egg collectors suggesting that nests contained eggs by at least 24 April. A nest in Sikkim province in northern India was at an elevation of around 850m, suspended at 3.1m in a small deciduous tree. The nest was neat, composed of grass stems, 'skeletonized leaves' and green moss. It was lined with fine grasses. But in contrast to other *Oriolus* nests, this one had stringed attachments to an upright branch behind the fork, making it very secure. The nest was made with fibres woven to a width of approximately 25mm; other material used included bamboo sheaths and plant tendrils. Another nest was described as not being suspended but partly wedged into and resting on a fork. Measurement of these and other nests revealed that the diameter was in the range 102–107mm.

The Silver Oriole, *O. mellianius*

The Silver Oriole *O.mellianius* was formerly known as Streseman's Maroon Oriole, and regarded as a race of *O. trailli*, despite it being silver where the Maroon Oriole's plumage is maroon. The Silver Oriole breeds in south-west China, where it is considered to be endangered because of destruction of suitable habitat. Other pressures being put on orioles and other bird species occur where a large amount of disturbance is caused by people harvesting bamboo shoots in spring. In the 1930s, the Silver Oriole was said to be very common, at least in Guanxi province, and fairly common in some wintering quarters at around the same time. More recently, it has been recorded in small numbers as far north as the southern mountains in the south of Sichuan province, where it breeds in evergreen broadleaf forests between 600 and 1,700m. It is these mature forests that are being fragmented, with the result that birds are less able to pair up and spread out into new territory. Nonetheless some birds have been noted in recent times in younger plantations at a higher density than in mature plantations, and it is thought that a rotational planting programme may help to re-establish a healthy population. The bird is usually found in pairs in the tops of tall trees from April/May to October. On arrival, noisy aerial courtship chases are performed above treetop level. The typical hammock nest is said to be quite large for the species' size, and is built by both the male and the female. One such nest was found in a local *Sorbus* tree species, at a height of 20m above the ground. Little is known yet of its behaviour or its diet, but we may assume that they are broadly similar to the behaviour and diet of other oriole species. Silver Oriole has been recorded taking nectar from the flowers of coral trees in Thailand. In October, the birds migrate south mainly to northern and eastern Thailand, and to a lesser extent western Cambodia, where they can sometimes be found in mixed flocks with other species around taller trees.

The Mimeta group

Sometimes referred to collectively as the 'Brown Orioles', most of the species in this group have been very little studied, and several are endangered as a result of loss of habitat. Those starred * are little-known birds. The main claim to fame of the Australasian members of this group is their propensity to mimic friarbirds. Sibley & Monroe (1993) recognise 16 species of friarbirds in the genus *Philemon*. Of these, three are mimicked by orioles.

The Brown Oriole* *O. szalayi* is found in the mangroves, scrub and forest edges of Papua New Guinea and the small islands of Waigeo, Salawati, Misool and Bantanto at its western end.

The Dusky-brown Oriole* *O. phaeochromus* is moderately common but is confined to Halmahera Island in the northern Moluccas.

The Grey-collared Oriole* *O. forsteni* is also relatively common, but is confined to Seram.

The Black-eared Oriole* *O. bouroensis* is confined to Buru Is. (southern Moluccas); the subspecies *O. b. decipiens* is confined to Tanimbar *Is*. (Lesser Sundas).

The Olive-brown Oriole* *O. melanotis* is found in the Lesser Sundas (Timor, Roti and Seman); the subspecies *O. m. finschi* is confined to one Wetar *Is*. (also part of the Lesser Sundas chain).

The **Olive-backed Oriole** *O. sagittatus* and the Green Oriole *O. flavocinctus* are the two species to be found in Australia, their range extending into Papua New Guinea. Sometimes referred to in Australia as the Green Thrush Oriole or Cedar Bird, *O. sagitattus* occurs from Broome in Western Australia through most of northern Australia, down the east coast and south-east to Adelaide throughout forested areas. While northern birds are sedentary, part of the population migrates south to breed between September and January. The Olive-backed Oriole also frequents the south-east of New Guinea. Although it does not mimic friarbirds, there is some evidence that it mimics the Green Figbird *Sphecotheres vieilloti*. Its plumage is a drab green that blends in well with foliage. The plumages of both sexes are very similar to that of the female figbird, and the orioles often feed with figbird flocks and may nest close to figbird breeding colonies. The Olive-backed Oriole's behaviour is secretive in the presence of figbirds, which ignore it, unlike the presence of other species, which usually elicits an aggressive response. While it feeds mostly on small fruit, the Olive-backed Oriole also takes insect food, sometimes hawking them in the air. Tom Tarrant (pers. comm.) believes that Olive-backed Orioles prey on nestlings, especially of honeyeaters, as he has seen the latter mob the orioles. The birds occur in drier woodlands than Green Oriole. There has been one instance of the Olive-backed Oriole species recorded as a vagrant on the Lesser Sundas islands of Wallacea in 1905 (Coates & Bishop 1997).

The Green Oriole *O. flavocinctus* is also known as the Australian Yellow, the Mulberry or the Yellow-bellied Oriole. It inhabits northern Australia and its offshore islands, as far east as Arnhem Land and the Northern Territory. The western limit is around King Sound in northern Western Australia to Broome. Across the Timor Sea it breeds on the small islands of Leti, Romang and Moa of the Lesser Sundas but not on Timor. In the New Guinea complex it is found along the southern parts and on the Aru islands. The bird likes hot, humid rainforests, mangroves, and monsoon and paper-bark forest, where it is one of many birds that feed on figs and other fruit. It is largely dominant over these other species. The plumage is dirty yellow, streaked brown (Frith 1976).

The Dark-throated Oriole *O. xanthonotus* is fairly widespread in evergreen forests and swamps up to 1,200m, from Burma down through the Malay peninsula, over much of Indonesia and Borneo and up to the south-western Philippine Islands. In the Sundiac Islands of Indonesia and parts of Thailand and Malaya, Birdlife

International has fears for its future because of forest destruction. In the Sundiacs it is thought the bird may disappear by 2010. Elsewhere its use of forest edge may help maintain a sufficient population for regeneration.

The Isabela Oriole *O. isabellae* is one of the two endemic species in the Philippines (where the Black-naped also occurs), the other being the **Philippine Oriole** *O. steerii.* The Isabela Oriole is a small oriole, at 20cm long. It is present only in about three localities in the Bataan province of the island of Luzon. In 1961, 11 Isabela Orioles were 'collected' in the Sierra Madre mountains. Since then, the forest in the mountainous areas has declined by 83% and so has the population of Isabela Orioles. At the actual site of the 1961 collecting expedition, illegal logging had destroyed most of the habitat left.

The bird has been recorded a few times in recent years: in 1994, it was recorded with a mixed flock of 12 other species (van der Linde 1996); in 2003, there were four records; and in 2004, a bird was netted and released (Weerd & Hutchinson 2005). It is considered Critically Endangered by BirdLife International.

There is a notable subspecies of the Philippine Oriole, called the White-lored Oriole (*O. s. albiloris*), which is confined to northern Luzon, and is also threatened by degradation of habitat.

A note on mimicry by Mimeta Orioles

In a genus in which the species are generally very colourful, the Mimeta (Australasian) group are very dull, being mostly brown or olive-green. However, this type of plumage closely mimics that of the friarbirds. This plumage imitation shows black-feathered areas similar to the friarbird's black facial skin. The mimicry even extends to behaviour, songs and calls.

Oriole species	*Friarbird model*	*Location*	*Comments*
Dusky-brown *O. phaeochromus*	Dusky *P. fuscicapillus*	Halmahera	virtually identical
Grey-collared *O. forsteni*	Grey-necked *P. subcorniculatus*	Seram	virtually identical
Black-eared *O. bouroensis*	Black-faced *P. moluccensis*	Buru, Tanimbar	also vocal mimics
Olive-brown *O. melanotis melanotis*	Helmeted *P. buceroides*	Timor	female, not male
Olive-brown *O. melanotis finschi*	Helmeted *P. buceroides*	Wetar	female, male less so

Table 2. Mimeta group mimics, plus models of the genus *Philemon.*

In Australia, however, where both Olive-backed Oriole *O. sagittatus* and Green Oriole *O. flavocinctus* overlap with the Helmeted Friarbird *Philemon buceroides*, there is no mimicry. The Brown Oriole itself is mimicked visually and vocally by the Streak-headed Honeyeater, even though this honeyeater is less than half the size of the oriole. Given that the friarbirds are generally larger than the oriole species that mimic them, and given that they occupy similar habitats and regularly feed together at the same flowering and fruiting trees, leading to disputes that the friarbirds generally win, it is not easy to see what advantage the orioles gain from their mimicry, unless it is some kind of 'big-brother' protection (Diamond 1982).

ICTERID ORIOLES

There is a group of birds in the New World called orioles, but they are unrelated to the *Oriolus* species. The Icteridae are divided into three tribes: Icterini, to which the icterid orioles belong; Dolichonychini or bobolinks; and Algelaiini, or New World blackbirds. The icterids not only resemble the Old World orioles, they have roughly adopted the equivalent ecological niche, have similar nesting habits, and they behave in similar ways with a similar diet – a prime example of convergent evolution.

There are 24 species of icterids, all found in North or South America. Some are restricted to particular Caribbean islands, the most endangered being the Montserrat Oriole, which had its available habitat slashed following a volcanic eruption in 1995. The Martinique Oriole is also considered endangered by the World Conservation Monitoring Centre. The decline is this case is not only because of habitat loss but also because of the spread of the brood-parasitic Shiny Cowbird *Molothrus bonariensis*, which first colonised the island in the 1940s. Another threat comes from the Carib Grackle *Quiscalus lugabris*, which has increased in numbers and which preys on both eggs and nestlings (Jaramillo & Burke 1999).

The only icterid oriole likely to be recorded as a vagrant to Europe is the Northern Oriole *Icterus galbula*, better known to us as the Baltimore Oriole (now considered the eastern race of Northern Oriole, with the former Bullock's Oriole, treated as the western race of the Northern Oriole). Baltimore Orioles have been recorded as vagrants to Europe on a number of occasions. They are mostly seen in Britain, but they have also been recorded in Greenland and Iceland (3+), Norway (1 male in May 1986) and Holland (1 female in October 1987). Only one has been recorded in Ireland (in 2006). Most Baltimore Orioles in Britain are casual sightings but increasingly they are being found to overwinter, as Table 3 shows.

In addition to Table 3, one found dead on 16 December 1968 in the English Midlands (Coventry) could easily have been frequenting gardens unnoticed until its demise. It seems likely, too, that those that were recorded during May in 1968 and 1970, and in Norway in 1986, were overwintering birds. There is also a record

Bird	*Location*	*From–to*	*Days*
First-year female	Haverfordwest, Wales	2 Jan 1989 to 23 Apr 1989	112
First-year male	Westcliff on Sea, Essex	2 Dec 1991 to 24 Mar 1992	112
First-year male	Headington, Oxfordshire	10 Dec 2003 to 16 Jan 2004	37

Table 3. Baltimore Oriole sightings in the UK.

from John O'Groats, the most northerly point of Scotland, of a male Baltimore Oriole in full summer plumage from 24–27 May 2007 (Nightingale & Dempsey 2007). The most in any one year was in 1967, when four were recorded, and the most concentrated series of records was six for the years 1966–1968 (Dymond *et al.* 1989).

Baltimore Orioles are generally in winter plumage on arrival in Europe. They are pale yellow-green with the breast sometimes showing an orange tinge. The birds are blackish on the head, back and scapulars but the amount varies, and the underparts, including the vent, are a dull yellow. Some males, by the time they arrive in Europe, may have started to develop a black throat. Autumn/winter females lack the tendency to black and are generally duller than equivalent-aged males. Both sexes' wings at this time are greyish-black with two dull white wing-bars. They are roughly starling-sized, chunky in appearance with a sharp, straight beak, which is a dull bluish grey. The legs and feet of both sexes at this time are greyish. When in breeding plumage in America the coloration is much brighter and more vivid.

These longer-stayers often visit garden bird tables, and when interested local residents realise what is frequenting their garden, they have put out oranges cut in half to encourage the bird to stay. This is a common practice in the United States, where orioles are frequent in gardens. The orioles have also been seen to feed on fat and niger seed put out for other garden residents in winter. It is virtually inconceivable that these ultra-rare vagrants make it back to the New World; none have been recorded in Europe in summer.

How these birds arrive in Europe in the first place is a matter of debate. Some are definitely ship-assisted, and most are recorded near to where ships pass or near ports. However, those found on the Scilly Islands and in other western counties of England could be brought over by Atlantic weather conditions, as are many other North American species in the autumn. That they can be ship-assisted is evidenced by the finding of a bird in the rigging of R.M.S *Mauretania* on 5 September 1964, two days and 16.50 degrees west, eastward bound from New York after a heavy swell created by a hurricane. It stayed aboard for some time until it disappeared, possibly as the result of being chased by a Red-shouldered Hawk, which itself left the ship near to the Fastnet Light off south-west of Ireland. The same liner had another visitor on 29 August 1965, when two specimens were aboard about 1,000km from land in company with a Yellow-breasted Chat, to be joined later by a Black-billed Cuckoo, Mourning Dove, a hummingbird species and a probable

Bobolink. On this occasion, all the birds had disappeared before the ship reached the Azores. In a systematic list of birds of probable American origin recorded on about 100 voyages across the northern Atlantic between 1961 and 1965, the Baltimore Oriole was seen on three occasions, involving four birds, which included a recorded by Nisbet of a bird in October 1962 which joined the *Mauretania* 3,000km east of New York, staying for several days pecking at limes and toast on open decks before finally leaving the ship some two hours' travelling time distant from the Irish coast. It was stated to be in good condition on departure (Durand 1972).

There is one other record of an Icterini species in Europe, and that is of a Black-vented Oriole *Icterus wagleri*: in Norway (Rogaland) from 10 July to 7 November 1975. In view of the fact that this is a non-migratory species, resident from north Mexico south to Nicaragua, it was almost certainly an escaped cage bird (Cramp & Simmons 1993).

CHAPTER TWO

The story of Lakenheath and the Golden Oriole Group

The East Anglian fens lie roughly within a line from Kings Lynn via the southern edge of the Wash to Peterborough in the north then south to near Huntingdon. The southern fen edge runs from there, following the course of the River Ouse to a point 3km north of Cambridge. From there the border runs north-east to near Newmarket and back irregularly to Kings Lynn. There are two main soil types: the northern section is mainly regarded as 'skirt', being loamy in character, whereas the southern half consists mainly of black peat. This is where trees need to be planted in shelter belts to break the strength of high winds, which can whip up the light peat topsoil into clouds of choking dust known as 'fen blows'. This in turn leads to land shrinkage, so much so that in many places the peat has disappeared down to its gault clay base and become less fertile. The peat soils are very valuable while they last because of the high-yielding crops they can produce. Apart from wheat, potatoes and sugar beet – the mainstay of agricultural production here – many vegetables are grown, including carrots, lettuce, onions and celery. Consequently there is a need to anchor the soil by means of shelter belts.

Ten thousand years ago the Fens did not exist, nor did Britain as we know it now, because it was still part of the European continent. The River Great Ouse was a tributary of the River Rhine. The area covered by the Fens today was once dense, tall forest, but at some time around 5000 BC the sea flooded in from the Wash area and eventually choked the trees, so that most were dead or dying before a great storm from the south-west laid them all flat. We know this because of the remains of trunks, known locally as 'bog oaks', some as long as 24m, that are now appearing as the peat shrinks. The water remained and gradually became less saline, until the reeds and meres became well established, creating a freshwater environment that was full of aquatic life but not very friendly for humans, as the myriad mosquitoes carried malaria. There is a good deal of evidence of the wildlife available to hunters during the Iron and Bronze Ages because of hunted-animal remains, which contain bones of Dalmatian Pelicans, White-tailed Eagles and Beavers, indicating a warmer climate favourable to orioles (Evans & Hodder 2006).

There were no organised attempts to drain the Fens until the Romans invaded Britain. Up to then all agricultural activity was confined to the fen edge and a few islands in the marsh, such as the Isle of Ely. The Romans did make some serious efforts to drain parts, but when they left the Anglo Saxons allowed it to return to its original state. It was not until the 1600s that King Charles I charged the Duke of Bedford to do the job properly. He, in turn, brought in Cornelius Vermuyden, a Dutch engineer, who established a system of main drains, channels and banks centred on the Ouse Washes, which still exists today, although this has gradually been improved (Gerrard 2003).

THE BLISS ESTATE AT LAKENHEATH

Until the East Anglian fens were drained in the 17th century, the Lakenheath area would have been a marshy bog with shallow, peat-bottomed pools with sandy, low, *Phragmites* reed-fringed islands or there would have been large areas of reeds where Bitterns would have bred, accompanied by Bearded Tits, Reed Warblers and Reed Buntings at least. Many waders such as Redshank and Snipe would have bred in large numbers. Grey Herons will certainly have bred, and Purple Heron is also a possibility. The smaller birds would be predated on mainly by Marsh Harriers in summer and Hen Harriers in winter. There would also have been a great number of passage waders. In winter, flooding would have diminished the land area and wildfowl would predominate, harried by Peregrines and, of course, by man.

Through all this the Little Ouse wound its way westwards to join up eventually with Cam/Great Ouse river system and on to the sea near Kings Lynn. Then in the 1600s, as Vermuyden set about draining the land, the Little Ouse along with all others, including the Rivers Lark and Wissey in the eastern fens and others to the west, were banked up and drainage ditches and dykes cut into the peat. Pumping engines were built to control water levels. As the peat began to shrink over time,

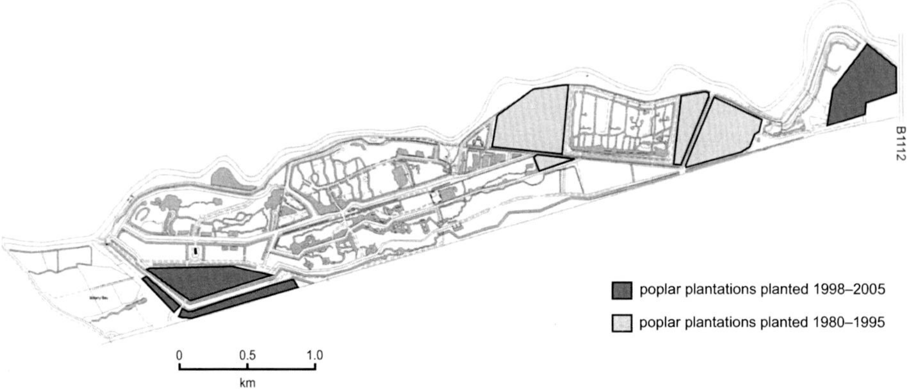

Figure 2. A plan of Lakenheath RSPB reserve showing the retention of poplar plantations and new planting at each end. Amended from map kindly supplied by Norman Sills of the RSPB.

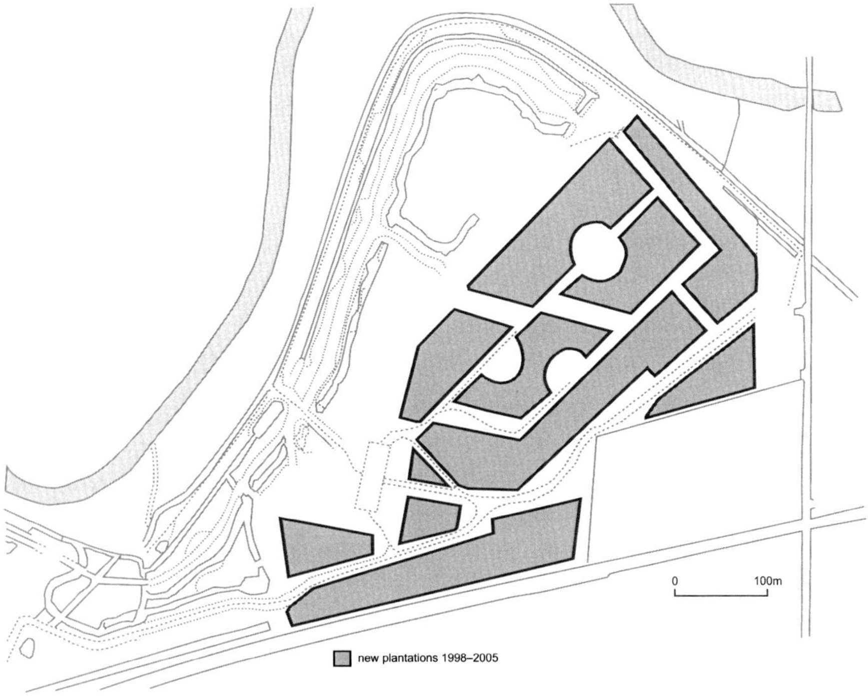

Figure 3. Enhanced detail of the layout of new poplar planting (1998–2005) at the eastern end of the reserve, showing the creation of open rides between the blocks. Amended from map kindly supplied by Norman Sills of the RSPB.

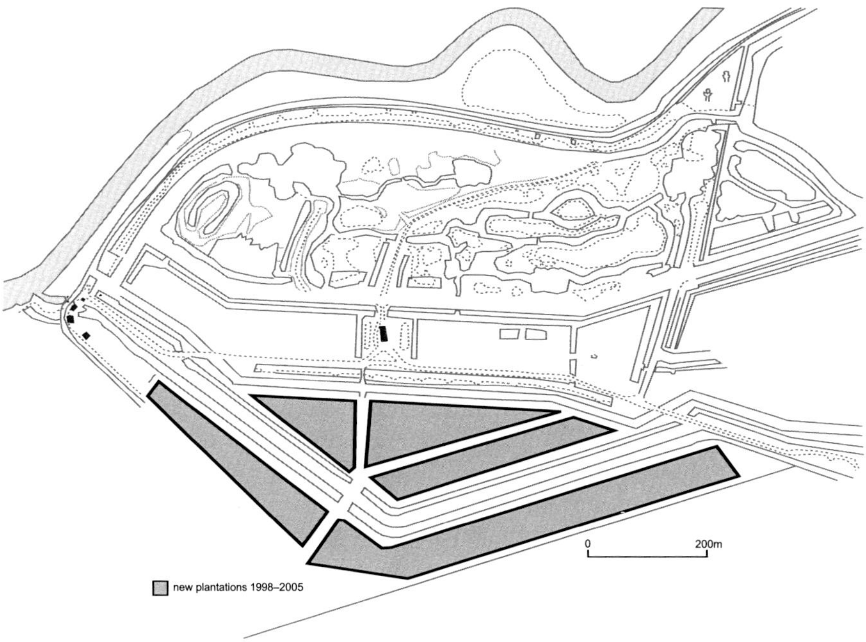

Figure 4. Layout of poplar planting at the western end of the reserve. Both the western and eastern ends of the reserve incorporated designs based on research by the Golden Oriole Group. Amended from map kindly supplied by Norman Sills of the RSPB.

rivers became higher than the surrounding land. As the fens dried out, the marsh habitat declined and people gradually took over an ever-larger acreage to farm in an increasingly intensive way. The peat gradually disappeared by natural shrinkage and, particularly in the months of March and April, strong winds would blow it away, creating black 'dust' clouds, filling ditches as it dispersed over other types of higher-land soil types. Another hazard that farmers in the Fens had uniquely to suffer was that dried-out peat would catch fire and smoulder away in a way that was difficult to extinguish. When eventually put out it could be seen that 'pitting' had occurred, and again the land would be rendered less productive.

The Lakenheath fen peat was of a particularly light character. Being shallow because it edged on the higher Breckland sand, attempts to try to bind the soil by sowing grass were not very successful. As a result the quality and value of the farmland here declined noticeably as the sandy island areas predominated and shrinkage progressed. During the 1800s a double-track railway was constructed from Ely to Norwich, which ran through the middle of the Bliss Estate. The estate itself comprised land in Norfolk Fen, Joist Fen, Shepherds Fen and New Fen, which in turn were intersected by Twelve Foot Drain, Furthest Drove and the intriguingly named Rightup Drove. Shepherds Farmhouse was the base and there was an outlying plantation in Grime Fen.

In World War II, any land, whatever the quality, was acquisitioned to produce food for the war effort. Later, following the catastrophic 1947 fen floods and the subsequent installation of a large cut-off channel (which was designed to prevent further flooding by diverting excess winter water around the edge of the fen where it met higher land), it was realised that a better use for this particular part of the fen could perhaps be the growing of poplar trees, intended ultimately for match production. Bryant & May, probably Britain's largest matchmaking company at the time, purchased 446.4ha, on which they progressively created blocks of a variety of hybrids. Planting commenced in the early 1950s, both here and on isolated farmland plots owned by other people nearby. As the trees matured, wandering Golden Orioles must have found them to their satisfaction and stayed to nest.

Breeding of Golden Orioles first occurred on the Bliss Estate in 1966, though it is thought from hearsay evidence that a pair had nested in a smaller plantation to the north-west the year before that. Local people found two nests in 1967, but for a few years the news was kept quiet, to prevent eggs being taken by collectors. More enlightened ornithologists, led by Mike Jeanes and the Suffolk Ornithologists' Group, were able, with the permission of the owners, to collect data from 1976 onwards, and by the 1980s breeding numbers at Bliss were at their highest. By this time word that a rare British breeding bird had set up a colony had got out to the birding fraternity (Tucker 1989), and the estate experienced several unfortunate incidents caused by overenthusiastic birders creating disturbances. Word also reached oologists, and several nests were robbed by these fanatical egg collectors. On one occasion two men were actually caught carrying a metal ladder.

That the place attracted attention is understandable as the aura was near to tropical when several male birds were singing and their song echoed through the trees in ventriloquial fashion. Other bird song which accompanied them was provided by the summer visitors – Turtle Dove, Common Cuckoo, Garden Warbler, Blackcap and Common Whitethroat – together with such residents as Blackbird, whose song is often mistaken at a distance for an oriole's. Song Thrushes found much sustenance from snails, mostly the colourful Grove Snail, found in large numbers in the humid atmosphere of the hemlock undergrowth, which often grows higher than a man. The snails also sometimes climbed in large numbers up tree trunks, and their shells could be found round many a stone 'anvil'. Treecreepers foraged up the same trunks, emitting their delicate calls, and Spotted Flycatchers hawked insects from branches near their nests in dead wind-snapped tree trunks. Even Hawfinches took up residence at one time. In one nearby plantation, Nightingales also joined in the dawn chorus. From all this the orioles song rained down from on high, giving the lie to the often-expressed opinion that commercial plantations were no good for wildlife. Marsh Harriers nested in a remaining marshland to the east called Botany Bay; to sit watching the harriers while listening to orioles singing behind was a unique experience.

In these dense ranks of poplar trees, as in many other fenland plantations, there is a substantial understorey, which includes extensive areas of hemlock, together with scattered elders and patches of alder carr. Alders have occasionally been selected by orioles for nesting, but being low, render the nests vulnerable.

The long lines of tall, trimmed tree trunks reach up to 28m at times, with a sun-dappled, dark green leaf canopy forming a sort of roof. The trees, planted at precise 5m intervals apart, together create an impression of a woodland cathedral, with birdsong echoing throughout in summer. In contrast, winter creates a bleak impression: the leafless stands are devoid of life, apart from woodpeckers searching out food on the many thousands of tree trunks. At this time, Hen Harriers are also recorded hunting between trees, and Barn, Long-eared and Tawny Owls can also occur.

Figure 5a shows the full extent of poplar at its peak. At this stage, the area held 14 breeding pairs of Golden Orioles. They were not nesting in the most southerly sections towards Stallode, as these were recent plantings, but the birds sometimes foraged there. In 1982, the owners decided they no longer wished to grow trees and the land was sold off to a farming company, Clopton Farms, who proceeded to return the estate to intensive arable farming, ironically at a time when government policy was beginning to adopt more environmentally friendly policies. Each block was felled over a period of about five to six years. First to go was the area in Shepherds Fen, apart from a triangular section near to the Stallode Wash (referred to later as the Retained Triangle). Shortly after, most of Joist Fen went, leaving two sections near to Botany Bay. In 1984, these last two sections were cut down, along with two other small compartments in Joist Fen near to the River Little Ouse. At this stage, with 90% of the poplars cut down, birds still nested in the remnant (as shown in Figure 5b). This is when the RSPB decided to intervene. The RSPB Regional Officer, Tony Prater, negotiated with Clopton Farms to allow a survey and

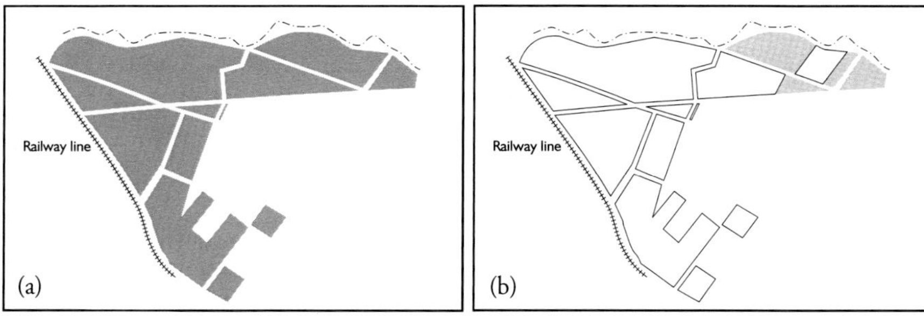

Figure 5. Bryant & May's poplar estate at Lakeneath at its maximum extent (a) when as many as 14 pairs of orioles may have bred; Lakenheath in 2000 at its minimum extent (b), shortly before its purchase by the RSPB, when the number of breeding pairs was down to two or three. The light grey areas show the remaining poplars at Lakenheath in 2000.

to retain140.5ha on a rolling planting programme. Even in the area surveyed, known as New Fen, some felling had started in the triangle near to the level crossing. Because we were interested to see if new varieties being developed in Belgium would be suitable for orioles, a trial plot of four or five new cultivars was planted in 1994–95. The use of the area by orioles following the changes in the early 1980s was investigated in the 1987 survey, details of which are given in Appendix A7.

That the estate would turn a full circle from being fen and marsh to drained arable to forestry and back to its approximate original condition started to happen when the RSPB showed interest in purchasing a large section as part of a national programme to reinstate reedy wetland in several parts of the country. Basically, this was due to a wish to create suitable breeding habitat for Bittern. What is now known as Lakenheath Fen RSPB reserve was bought in 1995 and consists of some 300ha lying between the railway line and the Little Ouse river. Norman Sills, who had acquired much knowledge of hydrology during his time as warden at the Titchwell Marsh RSPB coastal marsh reserve in north Norfolk, was brought in to develop the new reedbeds. The planned development of the Lakenheath Reserve also incorporated features to retain and expand the population of breeding orioles. The Golden Oriole Group was invited to advise on new plantations incorporating glades, drives and edges. The hope is that these new poplar stands will not only encourage oriole numbers to increase again but will also allow birdwatchers to see these elusive birds more easily.

THE GOLDEN ORIOLE GROUP: RESEARCH METHODS

Before linking up to form the Golden Oriole Group, three keen observers – Mike Jeanes (MJ), Malcolm Rains (MR) and Paul Mason (PM) – had been individually watching orioles in areas near to their home bases. The core site was then, and still is, the poplar plantations at Lakenheath, part of the Bliss Estate. Although living in Ipswich, MJ had been observing the birds at Lakenheath; MR was finding some in south-west Norfolk; and PM plotted the birds' spread into Cambridgeshire after first seeing a nest in 1982 (Mason 1993). In 1986, all three met at Lakenheath and discussed the merits of defining an area for the study of breeding orioles in the fenland basin. Seeing that poplars were the common factor in all cases, it was felt that the study area could be defined as one that encompassed all sites where poplars grew in reasonable amounts in fenland conditions. Following the investigation carried out by Jez Dagley of the RSPB we were able to map out the most likely area to concentrate on. Any site outside the core area could be brought in as necessary.

On studying the Ordnance Survey maps for the district, we discovered that all known breeding sites at that time were to be found on the Landranger 143 (Ely & Wisbech) map. This area is bounded by Downham Market east to the edge of the map as the northern boundary; from Downham Market, again, to the southern edge of the map via Ely as the western edge; and using the corner of the same map as the southern and eastern margins. Later the limits would be extended slightly further south. It is not possible to be more precise because of the need to respect confidentiality.

There is another small area on the western edge of the Fens next to Holme Fen NNR in what was then Huntingdonshire that contained birds for a year or two, but this happened before the Golden Oriole Group was formed (Clark 1996). This record is interesting in so far as Holme Fen consists, in fact, of 459ha of tall birch trees, flanked on its east and west sides by poplar plantations. Although birds were sighted in the birch forest, all confirmed breeding took place in the poplars. Nests at

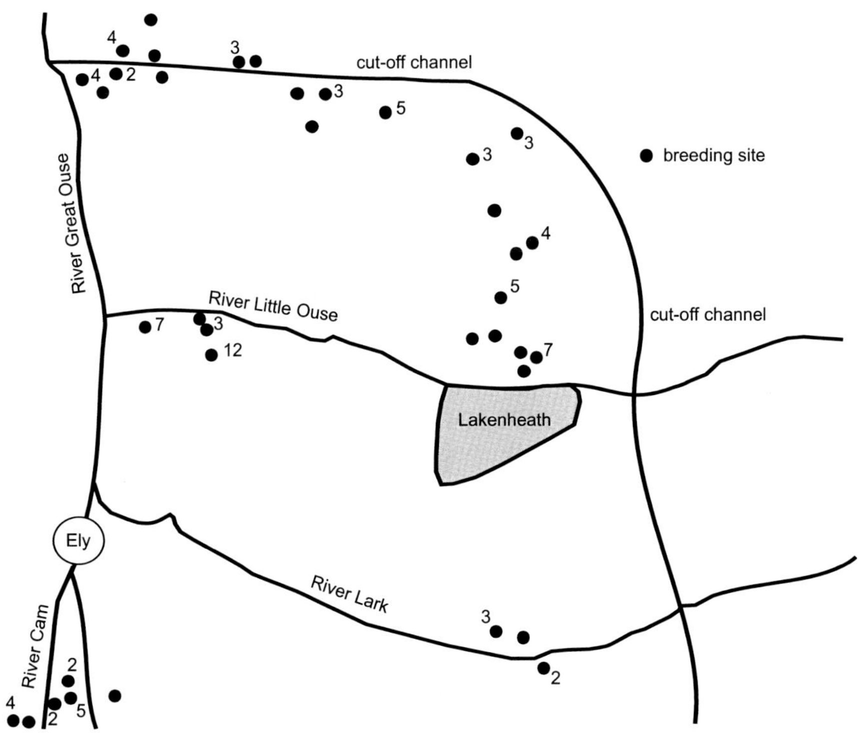

Figure 6. The Fenland Basin Study Area during the period of the Golden Oriole Group study, 1987–present (not to scale). Numbers are the number of times each site was used by orioles. All sites were used at least once but may have been used before the study period.

Holme Fen were recorded in 1982 (1 site), 1985 (1 site) and 1986 (4 sites), when the population in the main study area was at its peak. The fact that they bred in poplars rather than in birch suggests that they were an extension of the main population to the east, although orioles do breed extensively in birch at the north-eastern edge of their range in the Baltic States, Finland and north Poland, and, on rare occasions, in Britain.

The three original observers saw the value of combining their efforts by setting up the Golden Oriole Group, with a remit to carry out a systematic survey of breeding orioles in the fenland basin. A greater intensity and effort was needed to find the true extent of the birds' progress into suitable habitat. As most of the poplar plantations and shelter belts were on private property, the first task was to negotiate with the relevant landowners to obtain permission for access. In order to be able to approach any nests that were located, licences were obtained from the Nature Conservancy (later English Nature, and now part of Natural England).

Inevitably, the three members of the Golden Oriole Group could only devote a limited amount of time to survey work, given that they were all in full-time employment either as farmers (PM and MR) or in the service of a county council department (MJ). The situation was made worse by the immensely time-consuming task of locating nests in order to confirm breeding. Fortunately, through the good offices of the then RSPB Regional Officer, A. J. Tony Prater, a full-time fieldworker, Jez Dagley, was appointed for the 1987 season to carry out survey work. He was found a caravan to live in and placed in the middle of the Bliss Estate at Lakenheath, thus gaining unprecedented access to all 324ha of trees. This was by kind permission of the new owners, Clopton Farms Ltd, who had bought the estate from the matchmaking company, Bryant & May. Jez installed himself before the breeding season to survey as many of the sites as possible before the orioles arrived back from Africa. He and the other Golden Oriole Group members set to work using a standard survey form agreed beforehand. Each was allocated a sector of the agreed study area. Jez concentrated on core sites, although he later visited all of the others at various times to check on the presence or otherwise of birds at times when Golden Oriole Group members were unavailable. He also visited some places outside the Fens to assess the viability of plantations in river valleys never before looked at. Details of the 1987 survey and results are given in Appendix 7.

RESEARCH BY THE GOLDEN ORIOLE GROUP

Since 1987, detailed research has involved descriptions of the variations in habitat, including use by breeding orioles of various tree species. In the case of the hybrid poplar (*P. nigra* agg.), a knowledge of the range of cultivars and how to recognise them was acquired with help from various sources, principally the Forestry Commission. Lengthy watches at many nests were undertaken to ascertain feeding times and rates, bird behaviour, diet (particularly of nestlings), and other related behavioural matters.

Ringing was undertaken at 38 nests involving 106 young. Only one adult was ringed. Details of clutches were only estimated, except on a few occasions, at easily accessible nests. Difficulties of climbing very tall trees and a wish not to deter sitting birds in such vulnerable places precluded much of this activity. Details of nests were recorded, including orientation within the canopy on individual nest trees, height of nests from the ground, and distance from the trunk; even right- or left-handedness on given branches was noted (it has not been established whether this orientation has any significance). Proximity of nests to clearings and rides was ascertained, although many were in single or double belts. The ages of trees at first use was noted, and distance between nests was measured, sometimes during winter searches.

Biometric measurements of pulli were not taken because of the precarious situation the ringer was often in, although pullus size, condition and feather tract development were estimated. Estimates of fledgling numbers were generally made from family party sightings near to the nest after fledging. Diet information was obtained from faecal sacs, corpses and from photography, together with the help of researchers from Oxford University (grant-aided by TERF, the Environmental Research Fund). Their findings culminated in Foggo (1993) providing a comprehensive list of prey items available, though not all necessarily taken by orioles. A summary of Foggo's findings are given in Appendix 3.

Over the years not only did we find out how many breeding pairs there were and where, but we were also able to calculate the extent of territories, which was easier to assess in the more open situations away from the core at Lakenheath.

We did not climb to nests to ascertain egg numbers because not only were we concerned that the effort would entail the use of experienced climbers (which the average fieldworker is not), but we were worried that disturbance might cause nest desertion. In fact, the birds turned out to be unconcerned in general by human activity. Observations suggested that outlying pairs that saw few people were more wary than those close to greater human activity. In fact, some pairs nested quite close to houses and could be seen in gardens gathering flowers to decorate nests. The only time we did any egg counting was at a low nest, and even then a mirror on a pole was used. In any case, there is so much oological information to be found in papers published by European observers that we would not have added anything useful.

Because of the height of most nests, the best way to assess how many nestlings were being reared and fledged was by watching from the ground at the time of fledging. Unfortunately ringing could not be done at all nests because of the great height, the physical effort over a short, concentrated window of time and the vagaries of the weather. Moreover, some nests were inaccessible by virtue of being too far out from the trunk to be safe to reach. One difficult nest was only reached eventually by abseiling down from a strong limb above.

In addition, a great deal of information about diet has been collected over the years. As well as the study carried out by Andrew Foggo (1993), the findings of the Golden Oriole Group member Alistair Burn (AB), who is a professional entomologist, have been invaluable. He analysed faecal samples and was able to confirm that orioles also take beetles. His analysis of the stomach contents of a dead female oriole revealed that as well as the expected lepidopteran prey items, the bird had collected a shield bug and various hymenopteran species, including bumblebees (a supplement to Appendix 3 lists the species of moths trapped by AB in May and July 1992). To these, we may add the identification of prey items by those that spent a great deal of time patiently watching the comings and goings of parent birds satisfying the open gapes of their offspring. To this end photography was a great help. Behaviour at nest and the extent of the birds' travel to glean food were much easier to ascertain in more open sites, and it was found that the birds would forage at least 2km away at times and would sometimes take cleverly thought-out strategies to return to nests unseen by possible predators. A more detailed discussion of diet is given in Chapter 10.

OBSERVING ORIOLES

Orioles are notoriously secretive. For example, in 1994 at Site NOR06[1], no birds were recorded at all during the breeding season, and yet a nest was found after leaf-fall. Similarly, at NOR10[2], the only evidence of oriole presence was a male singing on 30 May, and yet a nest was found after leaf-fall. The weather can also make censusing very difficult. In 1995, the year of the national survey funded by English Nature and the RSPB, many sites were recorded as Category 3 (bird or birds present, but no evidence of breeding).

Fieldcraft is all important in trying to find evidence of breeding success. Once birds are located, preferably at the pairing and nest-building stage, watching back is attempted. It helps to have more than one observer so that triangulation can be achieved. When the general location of the nest site has been established, careful and often painful watching is required to pinpoint the actual nest. There is a painful condition known to oriole watchers as 'oriole neck', which results from tilting the head back for long periods of time in order to discern activity 30m or more up in dense canopy. To avoid oriole neck, GOG recorder MR went as far as lying on his back on a tarpaulin!

Even when the presence of a breeding pair has been confirmed, it is not always possible to locate a nest from the ground, even after spending an inordinate amount of time watching. One classic example of this was when a pair nested in a single-row farm shelter belt. Watching birds return with food was easy but like many species they did not fly directly to the nest carrying food. However, it was decided that the nest must be between trees 25 and 31. A walk under the trees produced nothing and the attempt to find the nest was ultimately abandoned. However, the ringer said he would go up 'on spec'. It was then that it became clear why we had not seen it: the placing of the nest had been so skilful that it could only be seen from certain angles.

Despite intense observer effort, many nests could not be pinpointed, although enough evidence was usually gathered to confirm breeding, e.g. sightings of birds carrying of nesting material. Even when a nest is pinpointed, it is often difficult to be sure of the number of chicks, except in those cases where ringing was carried out. In such cases, it is important to try to get sight of family parties once the young have fledged, in order to assess productivity. The analysis of the Golden Oriole Group data carried out by GOG member, Digby Milwright (DM), has provided an invaluable index to calculate maximum probable number of fledglings from a known minimum number.

Whenever we failed to locate a nest during the breeding period, we generally paid a visit to the site after leaf-fall, when any nests are likely to be visible, until, that is, they are blown down by autumn gales. These 'winter nests' were then collected and examined for signs of occupancy, such as feathers and other detritus. If a nest had been well worn or flattened at the open edge then it was a good indication that young had fledged. It was also not unknown to find a freshly dead nestling under a tree, but this occurred presumably when there was a large clutch of at least four well-fed birds that had fought for room in the nest. Such specimens were collected for gut analysis.

SITE COVERAGE

Sites are identified by a county prefix, CAM, SUF, NOR and HUN, plus a serial number to produce a coded name, e.g. CAM01. In a number of cases, for example in broad shelter belts and in the larger plantations, it has been necessary to subdivide; for example, what was originally NOR04 is now NOR04 [1], [2] and [3]. For the purposes of analysis, NOR04 is treated as three sites.

The number of surveyable sites has varied considerably in the period 1987–2005, rising from approximately 30 in 1987 to a peak of 85 in the mid-1990s (when a great effort was made to survey a lot of new sites, especially in Norfolk and Huntingdonshire, which seemed to have potential). As most of these additional sites proved to be unproductive, most were dropped by the year 2000. The number of sites available for survey has stabilised at around 55 in the last five years or so. Other causes of fluctuation include additions by splitting of sites, additions by reports of orioles in sites not previously known about, and removals of sites due to felling.

It is useful to have an idea of survey effort, i.e. the percentage of sites surveyed out of the total number of sites available. This information was not recorded for most of the study period, except for 2002 and 2004, for which the figures are 73% (37 out of a possible 51) and 64% (32 out of 56) respectively. As the number of surveyors has remained constant over the study period, and as no evidence is available to suggest that individual effort has varied significantly, it is probably safe to assume that in most years at least two-thirds to three-quarters of available sites were surveyed. Given that unsurveyed sites were regarded as unproductive for various reasons (a bird or birds recorded as present in the first year and not seen thereafter, deterioration of habitat, establishment of rookeries, etc.), one can say that to all intents and purposes, coverage has been close to 100%.

RECORDING OUTCOMES

Standard categories are used to record outcomes:

Category 1: Definitely bred
Category 2: Probably bred
Category 3: Bird or birds present on one or more occasions but no further information
Category 4: Site surveyed, no evidence of oriole presence
Category 0: Site not surveyed.

Clearly, Category 2 is not an absolute, so all records are discussed in detail by the GOG to determine the strength of evidence to justify promoting a probable to a definite. The third category is particularly frustrating, since the records are generally quite sparse, and quite often no more than a report from a farmworker or a gamekeeper with no further sightings (or sounds). The definition of Category 3 is of a bird or birds present 'at the right time and in the right place', i.e. in a previously occupied breeding territory (or one regarded as suitable for breeding) and within the breeding period. Very rarely does a Category 3 get promoted to a Category 2, but for the purposes of assessing the level of oriole activity in any given year, it is worthwhile to *assume* that every occupied site is a breeding site, i.e. to calculate the percentage of sites with birds present out of the total number of sites surveyed. Categories 4 and 0 are necessary in order to relate annual results to observer effort.

Every report of an oriole sighting, whether a spring arrival before birds are on territory or in an unusual location, is investigated. Gentle questioning of the informant often (but fortunately not always) reveals that the bird in question was not an oriole, usually because the bird was seen 'clinging to a telegraph pole' or 'feeding on ants on the lawn'. One particular example produced a moment of hubris for JA, who satisfied himself that the record was genuine on the basis of habitat, proximity to a known oriole area and plumage description, but also because the informant said that the bird perched on a telegraph wire, something, JA said confidently, that 'Green Woodpeckers never do'. Exactly one week later, as he was driving down a drove to a known oriole site, a Green Woodpecker flew out of the wood and perched for about 20 seconds on a wire before flying on. The moral of the story is where birds are concerned, never say never.

CHAPTER THREE

Habitat in Britain, Europe and North Africa

ORIOLES AND POPLARS

In British fenland poplar plantations, orioles are one of the few birds, apart from some corvids and some pigeon species like Turtle Dove, to use the trees for breeding. Green and Great Spotted Woodpeckers, Treecreepers and Spotted Flycatcher will also breed in suitable trees. It is reasonable to assume that the reason orioles use poplars almost exclusively in this country is that there is a good supply of the large prey items, particularly caterpillars, that they need to feed themselves and their young. Other foragers, such as tits, feed on much smaller prey, and the only serious competitor to the Golden Oriole is the Common Cuckoo. Cuckoos, of course, have only themselves to feed, having delegated the feeding of their young to host passerine species like Dunnock, Meadow Pipit and Reed Warbler.

The genus *Populus* consists of 30 separate species, all of which are from the northern temperate zones in Europe, North America and eastern Asia, mostly China. The most regularly found poplar species are White Poplar *Populus alba*, from central and southern Europe; Grey Poplar *P. canescens*, found in Europe and western Asia; Aspen *P. tremula*, native to Britain and found all through Europe and western Asia; the Chinese Necklace Poplar *P. lasiocarpa*, from China, mainly the south and west; and the Western Balsam Poplar *P. candicans*, endemic to Alaska and the west coast of North America as far south as California. Then, of course, comes the Black Poplar *P. nigra*, on which many of the commercial hybrids have been based. One such hybrid, robusta, has been for many years the preferred tree of orioles throughout the study area. The original Black Poplar is readily identifiable because the trunk, when full-grown, leans to one side before the sometimes massive branches ascend and arch out in a dome-like shape. The foliage is quite dense but the tree is not regarded as of commercial value since it is a slow-growing tree. It is widely distributed throughout Europe and eastwards to Kazakhstan in central Asia, growing in riverine situations and similar low-lying damp places. It is scarce in Britain because there has been no replanting programme until recently. Now, there is a move to restore Black Poplar to the countryside: conservation organisations are planting individual trees in appropriate sites, and the Tree Council has a programme of increased planting in natural positions.

Most primary species are endemic to certain climates and local conditions. For example *Populus sauveloens* is native to central and eastern Siberia. In Kazakhstan there are areas of Turanga which orioles, both Eurasian and *kundoo*, make use of. The Aspen, with a range stretching from the Arctic Circle in Scandinavia south to North Africa, and from the British Isles across most of Europe and north Asia to China and Japan, is one of the most widely distributed trees in the world. It is similar to white and grey poplars (the latter often found in Britain). While white and grey poplars are not regularly used by orioles, perhaps due to the silvery colour of their leaves contrasting with the straw colour of the nests, the aspen is used extensively, especially in eastern Europe, Russia and, notably, north-west Mongolia, probably because the yellow-green leaves tremble to give a dappling effect eminently suitable for orioles to hide their nests away in. Russia has the world's second largest area (20.6 million ha) of *P. tremula*, which is used commercially for many products. Because the species is self-generating through its root system of suckering, its management is easy and ongoing. In Siberia other natural poplars such as *Populus sauveolens*, *P. laurifolia* and *P. maximowoczii* amount to 962,700ha (Tsarev 2006), thus indicating a considerable amount of potential oriole habitat.

The poplar is probably the second most researched and frequently bred tree after the apple (Snell 1994), and there are an ever-growing number of hybrids being developed from *Populus nigra* using other parentage from *P. trichocarpa*, *P. deltoides* and *P. angulata*, known in the trade as 'Euro-American hybrids'. This cross-breeding work has been going on since the 18th century, mostly in Belgium but also in France and Holland.

Poplar can be a very fast-growing tree and it can be planted in areas where other main timber species, such as pines, cannot. However, whereas conifers will grow in more northerly climes up to the Arctic regions and on poor sandy soil, poplars will not do so satisfactorily. Poplars need wetter-based soil, such as bog, fen and river valleys, to flourish. So when the literature relates that orioles prefer damp habitats it is because a major tree species that they use happens to be there.

Not all commercially bred hybrid varieties are suitable for orioles to nest in. Some also have been discontinued by the trade as they are disease-prone, mainly to Canker *Xanthomanus populi*, or are liable to windsnap in exposed situations. To be attractive to breeding orioles, cultivars need to be early-leafing and to produce large leaves that will conceal the nests from predators and protect them from the heat of the sun. Among the cultivars that are less suitable for oriole nest placements because they do not meet the leafing criteria are serotina, eugenii and gelrica. In the Netherlands it was found that the hybrid cultivars zeeland, flevo donk and rap were suitable for breeding, whereas oxford, keppels, glorie and androscoggin were mainly used for foraging (Bijlsma 1995). Though giving long, straight trunks eminently suitable for timber purposes, the trees have a high, thin canopy with small leaves. In any case they have been superseded by more recently developed, faster-maturing varieties. Marylandica, which has a satisfactory canopy, has been discontinued as it has a propensity to develop a trunk with a kink, which renders it difficult to make planks from. Earlier clones took 25 years or more to mature but newer ones can be suitable for felling at around 17–18 years, thereby giving a better financial return.

The variety known as robusta has been around since about 1850, is widely used, is probably the strongest grower, and, as its name suggests, is robust in growth and development. It has been used in all sorts of situations, from commercial plantations, field shelter belts and small infill groups to just being a good decorative tree in parks and gardens. Despite the fact that it cannot regenerate, it does grow to quite an age before dying and is probably the most used cultivar, in Britain at least. Recently, two varieties of American balsam poplars have been developed: balsam spire and TT32, both of which produce early, dark-coloured leaves. These have been used by orioles, but there are not many plantations containing these cultivars. As time goes on, more are likely to be planted, as they form the basis for many of the newer, faster-growing timber varieties.

The more recently bred cultivars emanate from a Belgian arboreal research station at IBW Geraardsbergen. To investigate the new cultivars' suitability for oriole nesting, some GOG members visited the research station where the existing cultivars and several others not yet released for sale were being grown. Those inspected were boelare, beaupré, gibecq, ghoy, gaver, primo and trichobel. It appears that the first four named are the fastest growing, with boelare and beaupré likely to give the best financial return. Other factors have to be taken into account, specifically resistance to diseases like Canker and to *Melampsora* rust, and susceptibility to windsnap. Boelare and beaupré do have a windsnap problem, but the risk is reduced if they are surrounded by ghoy in large plantations (Tabbush & Beaton 1998).

Cultivar	*Leafing time*	*Leaf size*	*Disease prone?*	*Prone to windsnap?*	*Growth*	*Suitable for nesting orioles?*
robusta	early	large	moderately resistant	little	moderate	yes
serotina	late	small	–	little	poor	no
heidimij	late	small	–	moderate	poor	no
marylandica	early	large	–	little	moderate	yes
TT23	early	large	moderately resistant	little	fast	yes
beaupré	early	large	resistant	yes	very fast	yes
bolaire	early	large	resistant	yes	very fast	yes
ghoy	early	large	moderately resistant	little	fast	yes

Table 4. Features of cultivars in the Golden Oriole Group Study area.

To establish the suitability of cultivars for breeding orioles, the Golden Oriole Group researched the foliage structure, and particularly early-leafing and large-leafing propensities. We concluded that boelare, beaupré and ghoy (even though leaf size was slightly smaller in ghoy) were satisfactory. For this reason, we recommend those varieties, together with the old stalwart robusta, which is still the best for shelter-belt use. For the record, the Lombardy poplar (*Populus nigra italica*) is highly decorative but is not suitable for nesting because the branches reach upwards at too acute an angle, though they are used for foraging.

Table 4 lists cultivars researched by the Golden Oriole Group in its study area. There are others that may be suitable for orioles, including new ones that have come on to the market since this research was carried out.

In the UK there exists a voluntary trade body known as the Poplar Forum, which issues periodical newsletters where much up-to-date information can be obtained regarding growers' experience in various aspects of the trade. It contains information on changes that may occur in schemes such as the Farm Woodland Premium, and other information on uses for the timber. Another factor that may influence attitudes to poplar planting is the introduction by the EU of Single Farm Payments, replacing production subsidies. Many farmers are not sure how this will affect crop and livestock production, so tree planting may not be considered worthwhile unless forestry is further encouraged by government.

In order to encourage the planting of poplars in Britain, landowners and farmers need to be convinced of their value. In Holland on the reclaimed polder areas, and in Belgium, northern Germany and France, poplar planting has been actively promoted by government, because there and elsewhere on mainland Europe the timber has a wide variety of uses (Table 5).

Form	*Uses*
Biomass	energy production, fibre and chemical products
Peeled veneer	plywood and glulam products, food and fruit packaging, industrial packaging
First-quality timber	furniture, planking, instruments
Second-quality timber	pallets and sheds

Table 5. Uses of Poplar. From Storrs, 1994.

In France in particular, poplar is used as sawn timber or veneer for plywood. It is also used for interior joinery, frameworks and moulding, as well as for block boards. Furthermore, poplar is also suitable for crates, and light and heavy packaging, as well as pallets, showing that there is great potential in the UK if only the timber trade could be persuaded.

In the 17th and 18th centuries, poplar wood was used for making harpsichords. In the last century, poplar was grown in Britain for match production, but this has largely died off as a result of the invention of new lighting methods such as disposable lighters. This is why Bryant & May, famed for their matches (but who also made wooden baskets and packaging), decided to discontinue operations at Lakenheath and at other plantations in Bedfordshire and Herefordshire.

Storrs (1994) proposed that there should be an attempt to overcome user prejudice to poplar in the UK by publicising its extensive use overseas. Greater use for building construction purposes such as rafters, roof trusses and floorboards should be encouraged. It is a relatively odourless wood and takes stain and paint easily, so furniture makers could use it. Its use for making coffins could also be encouraged. Unfortunately, this is a Catch 22 situation; there are not yet enough mature trees for any timber yard to invest heavily in handling poplar, and there are not sufficient customers asking for this wood because continuous supply cannot yet be maintained. According to Storrs, if growers go for high quality, produce a standard and graded product, provide reliability and continuity, and get closer to the value-added market, they should succeed.

Since 2000, a certain amount of new planting has taken place, particularly at Lakenheath. It is to be hoped that the decline in oriole breeding numbers in the Fens will be reversed when the newer trees reach a suitable size. The RSPB asked the Golden Oriole Group for advice in designing the layout and cultivar content of the new plantations at Lakenheath. These will not be the usual rectangular commercial blocks but have been designed to maximise edges. This should give the birds more choice of nest placings and should allow visitors to the reserve a better chance of seeing the birds. The cultivars used should age at different rates, thereby creating their own felling rotation. Moreover, the RSPB will be able not only to keep the birds on site but will also derive income from the sale of mature wood, which could finance an ongoing replanting programme.

Nearby landowners are also showing some interest in planting poplars again with grant aid. Although availability and size of grants and conditions attached to them make decisions difficult, we were able to provide one local estate with an action plan, recommending a felling and replanting regime like the one at Lakenheath. In another initiative, the Woodland Trust, in cooperation with Haddenham Conservation Society, a parish-based environmental group in Cambridgeshire, designed and planted up a 3.6ha plot, known as Nineacre Wood. This now contains a mixture of hybrid poplars and other deciduous trees to try and attract orioles, which have bred to the north and south of the site. This is the only time the Woodland Trust has agreed to plant non-regenerative trees, with a similar felling and replanting scheme as at Lakenheath. Some 60km to the north of the main study area, a large commercial plantation of all the new cultivars has been installed at Wormegay. Here, too, the Golden Oriole Group was asked to advise on the layout. One lives in hope that the birds find and make use of this place when the trees are mature enough.

WHY DO ORIOLES BREED WHERE THEY DO?

A question often asked is why Golden Orioles nest in poplar trees at a height within trees that often reaches 27–30m when other species do not. The answer is that this represents an ecological nesting niche, at least in Britain and much of northern Europe, taken up by no other passerine species, apart from corvids. Whether the same is true in more southerly European countries is uncertain, but Fieldfares are recorded as using the same trees in Germany, albeit placing their nests close to the trunk (Norman 1994).

Feige (1986) suggests that where water creates openings in woodland, a local climate is set up, which encourages plant growth and a consequent improvement in the supply of insect food. Where our species breeds at higher altitudes than most it is mostly in river valleys in mountainous countryside up to about 2,000m (Fehringer 1922; Heyder 1952). Feige also believes that orioles have a tendency to nest in the tree species in which they hatched, even within mixed woodland. In England this may be why fenland birds breed almost exclusively in poplars, whilst in Kent and Essex they frequented other tree species, even where poplar was available. It may also account for the use of silver birch by a pair that bred at Wicken Fen in 1999.

OTHER TREE SPECIES USED BY BREEDING ORIOLES

Elsewhere in the Golden Oriole's breeding range, poplars are extensively used, but not exclusively so. The only countries where orioles have not been found breeding in poplars are in the eastern Mediterranean states of Israel and Lebanon, in Saudi Arabia, in the Baltic states of Lithuania, Estonia and Latvia, and in Finland. Russia has a great variety of habitats, including poplar and its near cousin, aspen. In Central Siberia as far as Lake Baikal, the birds use an eastern variety of primary poplar, *P.opulus sauveloens*, and *P. laurifolia* near the border with China and Mongolia. Between the taiga regions and the Stavropol region, habitat gradually changes and includes the use of field shelter belts, ranging from poplar to willow to birch, and becomes more and more riverine in essence. Poplar is also used in Azerbaijan, but a large percentage of breeding is to be found in commercial pistachio nut plantations (E. Sultanov, *in litt.*). In north-west Europe, the birds use poplar almost exclusively, mostly around the coastal North Sea regions. In France, the orioles' choice of tree species becomes more catholic the further south one goes, with poplar predominating in Normandy, Picardy and the Pas de Calais.

Oaks of the genus *Quercus* are also used by orioles. Sessile Oak *Quercus petraea* is native to much of Europe. Being a large, tall tree from 20 to 40m tall, it is similar to the traditional Pedunculate or English Oak *Q. robur*, with which it overlaps extensively in range. Sessile is more frequent than English Oak in upland areas with higher rainfall, and also with lighter soils with better drainage. Such conditions are found more in central and southern Europe, and in these areas the Golden Oriole uses Sessile Oaks more readily than further north, probably because of their height, which makes then similar to tall poplars. Downy or Pubescent Oak *Q. pubescens* is native to southern Europe and south-west Asia. It is also found in France and parts of central Europe such as the Czech Republic, Slovakia and Austria. It is a medium-sized tree, growing up to 20m in height. Large trees develop very thick, whitish bark cracked into deep furrows. It was in these trees that orioles were observed nesting during a visit to Bohemia (Trebonsko region) by PM and David Hopkins (DH) in 2006. Pedunculate Oak is native to most of Europe, Asia Minor, the Caucasus and parts of North Africa. At an ancient woodland nature reserve in south-east England, Blean Woods, all the Pedunculate Oak in high forest is of the same age, with trees around 150 years old. Hornbeam coppice also occurs, interspersed with oak stands and introduced sweet chestnut. It is at this wood that orioles nested for three to four years in the 1980s in the sweet chestnut coppice understorey. In Britain this tree mainly grows in open forest or parkland situations, where it develops a large dome-shaped canopy, often the result of pollarding. This could be why it has not been adopted by orioles in this country. Cork Oak *Quercus suber* is a medium-sized evergreen oak, native to south-west Europe and north-west Africa, and widely cultivated in Spain, Portugal, Algeria, France, Italy and Tunisia. It provides a very suitable nesting habitat for orioles and other species.

Species of birch in the genus *Betula* are also used by orioles. Silver Birch is widespread, although in southern Europe it is only found at higher altitudes. Its range extends into south-west Asia in the mountains of northern Turkey and the Caucasus. The closely related Siberian Silver Birch in northern Asia and the Sichuan Birch of central Asia are also treated as varieties of Silver Birch by some botanists. Silver Birch is distinguished from the Downy Birch, the other common

European birch, in having hairless, warty shoots (hairy and without warts in Downy Birch) and whiter bark, often with scattered black fissures (greyer and less fissured in Downy Birch). The two have subtly different habitat requirements, with Silver Birch found mainly on dry, sandy soils, and Downy Birch more common on wet, poorly drained sites such as clay soils and peaty bogs. It is in such conditions from the Baltic States, eastern Germany, Poland, Finland and through Russia and Siberia that orioles choose to nest, and throughout this range Silver Birch is used extensively by orioles. Only once have they been recorded as nesting in Silver Birch in England and that was atypically in Wicken Fen (Cambridgeshire) in a group of four such trees surrounded by lower bushy sallows.

A number of other species are used by orioles either for nesting or for foraging, or both. Alder is the common name of a genus of flowering plants belonging to the birch family Betulaceae. The genus (*Alnus*) comprises about 30 species of trees and shrubs, a few reaching large size, distributed throughout the Holarctic. The best-known species is the Black Alder, native to most of Europe and widely introduced elsewhere. Orioles have often been recorded nesting in these trees, especially in Germany and central Europe. In Britain, there are three records of orioles breeding in alder carr but under a poplar canopy at Lakenheath.

European Ash is native to most of Europe, with the exception of northern Scandinavia and southern Mediterranean Europe. Like willow it appears not to offer great camouflage opportunities for nesting orioles. The leaves of the European Ash are often among the last to open in spring, and the first to fall in autumn if an early frost strikes. As such it is less likely to host an oriole nest. Notwithstanding that, this tree is recorded as being used by orioles in moderate amounts. In Denmark and Sweden it is said to be used within a woodland otherwise containing only beeches.

There are some records of orioles using conifers. Use by orioles of Stone Pine, native to southern Europe, primarily the Iberian peninsula, has been recorded quite extensively in Portugal (B. Martin, pers. comm.; H. J. da Silva Costa, *in litt.*). This species also occurs in the Altai (Siberia), where orioles occur. Little information is available on the use of other conifers by orioles, except for references to Scots Pine being used occasionally in north Russia and Siberia.

Among more exotic tree species used is the *Robinia pseudoacacia* or False Acacia. *Robinia* species are used as food plants by the larvae of some Lepidoptera species including Brown-tail, Buff-tip, Engrailed and Giant Leopard Moth and so would be a good source of food for the birds. In several countries bordering the Black Sea, such as Romania, Bulgaria, Hungary and Moldova, these trees have run wild and orioles have taken to using them quite extensively. Another unusual species is the Oriental Sweet Gum, endemic to south-western Turkey. The range of this species has declined by 80% since 1949. Nowadays it occurs only in scattered patches. Its attraction for orioles is probably that it has large leaves and that it forms dense stands, usually in wet areas, with open areas between the blocks of trees, so there are plenty of nesting and feeding opportunities. It is also one of the three taller species found in south-west Turkey, the others being poplar and willow (G. Welsh, *in litt.*).

Another exotic is Eucalyptus, used around the Mediterranean countries and in the Arabian peninsula. At first sight, its leaf structure does not seem good enough to hide nests in, but it appears that, in Israel at least, the trees used are very high and around 100 years old (Y. Perlman, *in litt.*). It might be, therefore, that the birds consider height to be this tree's main advantage. In Portugal Eucalyptus is said to be used as long as there are other tree species nearby (H. J. da Silva Costa, *in litt.*).

Trees that seem to be less attractive to orioles include willows, elms, Ash, beeches and Hornbeam. Where these trees are used, it seems to be because where breeding densities are high, and so orioles without territories are obliged to move into what is effectively suboptimal habitat. This seems to have occurred in Belarus, where orioles moved into Crack Willow along riverbanks (Norman Sills, pers. comm.). Other trees can be seen to have certain disadvantages. Lime, for example, has not been recorded as being used, even though it is large-leafed and looks suitable. Lime may be unsuitable because its leaves attract a large amount of honeydew, which would drop on to the nestlings' plumage and affect the efficiency of the feathers. Willows, although in the same family as poplar and certainly found in wet soil conditions, generally have a leaf structure that would not satisfactorily hide the presence of a nest placed on a branch about two-thirds out from the tree trunk. Leaves are narrow, mostly vertically hung and not dense, thus allowing light to permeate through the canopy. English and Wych Elm leaves are also small, and the

Figure 7. An elm bough in winter. The structure of an elm makes it unsuitable for orioles to nest in. The leafy area is generally too exposed to the elements in summer, unlike poplar, for example. Paul Mason.

branching structure of the mature tree does not appear to be suitable for the way orioles construct their nests; the sub-branches are generally at too wide an angle for the bird to satisfactorily weave the nest materials across. Where the structure might be suitable, it is far too near the outside to be of use because the nest would be much more exposed. Ash has a leaf structure of 10–11 small, compound leaves on one stem towards the end of any given branch, which, as in the willow, allows space for light, thereby making it incapable of hiding nests. However, in this regard Pascal Etienne (pers. comm.) has noted Ash being used in the Marquenterre, and there are records from Switzerland and Portugal, and from Denmark, where it is used when present as a copse within beech woods (G. Fahlberg, *in litt.*). Observations in France (PM) suggest that beech woods are unsuitable as they form a very dense canopy without many clearings.

Everything we have discussed thus far relates to breeding territories. As the fledglings grow, the parents take them to habitats very different from the natal area in order to find food. At this stage, before the family finally breaks up, any suitable foraging area can be used, especially as the birds start to add fruit to their diet. On migration, a much wider range of habitats are used: cliffs, reed beds near to shorelines, vinyards, olive orchards, etc. In deserts, oases will certainly be taken advantage of.

BRITAIN

THE FENLAND STUDY AREA

By 'the Fenland study area' we mean the area that has been the subject of research by the Golden Oriole Group, i.e. the fenland basin of East Anglia. In the study area, birds have bred at 40 different sites at least once. Within that number some sites were divided further, as at Lakenheath, where at its peak 14 different singing males were recorded (assessed by triangulation involving several people), though this did not necessarily mean there were 14 nests. When this site was spread over 350ha consisting mainly of very tall poplar trees, it was very difficult to find every nest in order to provide an accurate estimate of the number of breeding pairs. The occupancy of a few of the more isolated sites has been very variable, with some being used only once and others occupied for seven to 10 years without a break. Some would be unused for a year or two before being re-occupied by a new pair. These sites were almost exclusively stands of poplar hybrids grown for commercial forestry use. They were either plantations to provide wood for the match trade or farm shelter belts (one to two trees wide) around fen fields, where the topsoil was apt to blow away in strong winds.

When the match trade declined during the late 1970s and early 1980s, the plantations were either cut down or left until, being non-regenerative, they collapsed, by which time orioles found them unsatisfactory. The market having been downgraded, landowners discontinued planting, so age rotation was disturbed. In the late 1980s some replanting was resumed, encouraged by government forestry grant aid, including fresh planting by the RSPB at their new Lakenheath reserve.

Within the plantations, shelter belts, and even in isolated clumps of poplar, it was found that only certain clones were used for nesting, whereas foraging took place in all types. Those selected for nesting were of the early-leafing, large-leafed varieties like robusta, balsamifera and heidemij. Newer varieties like beaupré, bolaire and ghoy have since been found to be attractive to orioles, and as they also grow more quickly and produce better-quality wood, they are now the cultivars being set. It was noted that if there were only a few robusta in a plantation of serotina or gelrica – two of the unsuitable types – then it was invariably the robusta that was used (Dagley 1987, 1994).

In the larger plantations the birds invariably chose to place nests in trees near a glade, perhaps caused by fallen trees, or next to a ride within the block. Then a tree two to three rows in from the opening was selected. Single- or double-row shelter

Lakenheath

Nest site	*Area (ha)*	*No. trees*	*Nest ht (m)*	*Cultivar*	*Fledglings*
1	33.84	5414	26	heidimij	1
2	21.63	3469	23	robusta	failed
3	14.20	2272	21	heidimij	failed
4	12.10	1936	25	robusta	1
5	26.16	4185	21	heidimij	failed
6	4.87	542	8	robusta	4
7	2.12	616	25	robusta	failed
8	2.5	294	23	robusta	1

Elsewhere in Fenland basin

Nest site	*Area (ha)*	*No. trees*	*Nest ht (m)*	*Cultivar*	*Fledglings*
9	4.5	–	18	robusta	failed
10	4.9	–	12	robusta	1
11	5.38	–	13.5	robusta	failed
12	3.69	–	11	eugenii	1
13	–	–	10	robusta	2
14	(belt)	–	9	robusta	3
15	–	800	6–10	robusta	failed
16	1.5	153	–	Balsam	failed
17	6.0	–	16	robusta?	failed
18	10.0	1910	7.5	marylandica	3

Table 6. Results of the breeding survey of Golden Orioles in the Fenland Basin in 1987. Nest-site numbering follows Dagley, 1987.

belts were also favoured, provided there was a sufficiency of other trees in which to forage. Size of territory in these conditions was difficult to assess because foraging birds would need to fly over much open ground. We were able, however, to arrive at an estimate of 400 as the minimum number of trees needed by any individual breeding pair. Very occasionally an isolated block of as few as 20 to 30 trees held a nest, which meant that the adults had to travel some distance to collect food.

In the study area, on only six occasions did orioles nest in trees other than poplar: four times in alder carr under a poplar canopy; once in an apple tree 1.2m from the ground, although the orchard was surrounded by poplar belts (R. Hoblyn, pers. comm.); and once in a Silver Birch, where the nearest poplars were about 1km away. This was at Wicken Fen, Cambridgeshire, in 1999.

Elsewhere in England, particularly in Kent and parts of Essex, Sweet Chestnut coppice under an oak canopy was used during the 1980s. In 1995 in Yorkshire, poplars held a nest, with a similar record in Lancashire some years earlier. Orioles also nested in Fife (Scotland) in 1974. The observer was unable to ascertain which tree the nest was in, but he subsequently viewed a family party with recently fledged young in conifers. There were no poplars in the vicinity, and nearby alders and birch drew a blank when searched. Being so far north, there is also a possibility that White Birch was used, a tree species widely used in Nordic countries and throughout Russia. This pair may have been migration-drift birds bound for those parts.

A search through historical British records reveals a variety of tree species used, including oak and elm. Birds nested in Hertfordshire in 1970 and 1971 in habitat described as 'parkland'. The 1971 record was of a family party, so actual nest site details are not recorded. Old county records and other literature also provide some interesting records, including an account from 1870 of a nest at a place near Diss in Norfolk. The nest was built in creeper plants over a trellis frame surrounding a pheasant-rearing pen, but no further details of the general habitat are recorded (Bannerman & Lodge 1953). Another nest was located in Holly at a site near

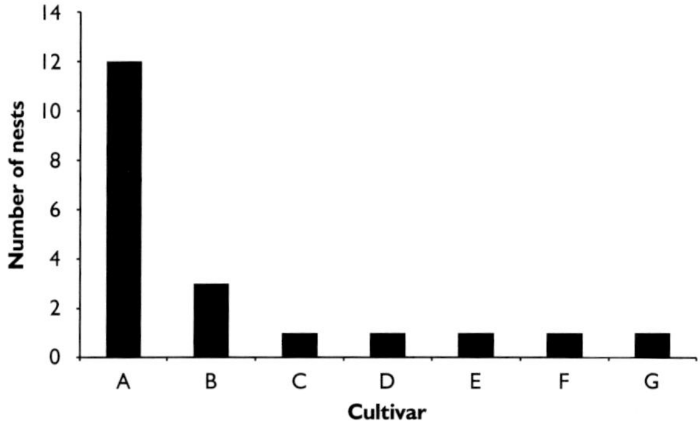

Figure 8. Use of poplar cultivars by breeding Golden Orioles in the Fenland Basin in 1987. A = robusta, B = heidermij, C = balsamifera, D = marylandica, E = TT32, F = eugenii, G = serotina. After Dagley, 1987.

Saffron Walden in Essex. Despite diligent searching of the literature from the birds' entire breeding range, this is the only record we can find of Holly being used to nest in. Historically there have been several records where no note of tree species used was made, but one in Lancashire, in 1959, of a nest in a fallen Larch is problematic. Apart from it being in a most unusual part of Britain, which is quite possible, there is a query of correct recognition of the tree species, especially as Larch has never been recorded before or since, and that it was a fallen tree and the nest was only about '4–5 feet' (1.2–1.5m) above the ground. It was found by a young boy of 15, who had collected an egg (one of five), which appeared fresh, so the bird species was not in question, but the author of the note questioned him closely about it and the boy, who had not seen orioles before, gave a good account (Barnes 1960).

CONTINENTAL EUROPE AND NORTH AFRICA

It is clear that one of the reasons that the Golden Oriole is so successful in the main parts of the European mainland and western Asia is because its habitat requirements are quite catholic. The warmer the climate and the more southerly that breeding occurs, the wider the variety of tree species used. It seems to be only on the north-west fringes of their breeding range that they restrict themselves to only one or two tree species

The bird breeds throughout most of Europe and eastwards through Russia and other former republics of the USSR, as well as in Morocco, Algeria, Asia Minor, the countries around the Black Sea and northern Iran. It is not surprising, therefore, that its range of habitats varies considerably throughout this range. Where it is common, it becomes a garden bird, as well as inhabiting parks, cemeteries, orchards, cork oak plantations, conifer plantation edges and small groups of trees and bushes in open farmland. A common factor to most of these situations is the proximity of water, in the form of rivers, canals or large drains. If there is a sufficiency of trees then streets, avenues and even peat-cutting areas are known to have been used (Melde & Melde 1977) Breeding has also been observed exceptionally in low bushes in steppe where no trees grew (Johansen 1944). In France, at the Marquenterre nature reserve at the mouth of the River Somme, orioles can be found breeding mainly in birch trees, and foraging in Sea Buckthorn bushes in the sand dunes (Etienne 1985). In north-west Russia and the Karelia region of Finland, there are vast areas of birch forest, which by nature is an open type of woodland, of which a large proportion of the estimated 2,000 pairs in Finland will make use. Information from Lithuania and Latvia describes orioles as being widespread in boggy birch and alder forests, similar to the habitat in which they are found in northern Poland. Estonian numbers are said to have declined rapidly in recent years, but the reason for this is unclear.

In Eastern European countries as Bulgaria, orioles are very common, using a wide variety of trees, including Aspen and willows. Hungary has had several poplar-planting programmes on land where natural forest had been cut down, a very mixed blessing. In southern France, oak is often reported by birders as being the host tree, but in northern France, at least, poplars are grown on a commercial basis in river valleys and orioles make good use of them. All observers note that orioles do not use dense woodland unless there is an opening, a glade or a swampy area within.

While most Eurasian Golden Orioles breed at altitudes from sea level to about 500m, there are notable exceptions, such as those breeding at 1,000m in Romania and the Italian Apennines, and at 1,500m on the foothills of Mount Etna in Sicily. Elsewhere around the Mediterranean, 1,000m is attained in the Pyrenees, and in Morocco up to 1,400m. In the easternmost part of their range, the Altai Mountains of central Asia, breeding has been recorded at 1,100m (J. Shergalin, *in litt.*). The Indian Golden Oriole is disposed to breeding at even higher altitudes and has been found nesting at 2,200m in Turkestan and at 2,700m in the western part of the Pamir mountain range bordering Afghanistan (Ivanov & Pogrebnoi 2002).

Morocco and Algeria

In Morocco, the main habitat where most orioles breed is described by Thevenot *et al.* (2004) as being in very dense, 'half wet' oak forest, under which there is little or no vegetation. This is mainly in the Haut-Atlas region, and breeding densities here are assessed at one to two pairs per 10ha, up to an elevation of 1,800m. Breeding was once recorded at 2,500m in the upper reaches of Oued Amizmiz. More recent information states that the birds breed quite commonly in cork oak on the plains and foothills up to 1,200m. Poplar is used around lakes and riverine situations, olive groves and in irrigated orchards. They can sometimes be found in parks and gardens in towns like Tangier, Rabat and Marrakesh. Meade-Waldo also says that they are rare or absent in coniferous woods. According to Cody (1987), they breed at up to 1,400m as far south as Beni Mettal. The same probably applies in Algeria but little information is available for this country. There is also one record (Isenmann & Moali 1987) of a nest in Holm Oak containing four eggs on 1 June at Ouarsenis, Algeria (year not recorded), but most birds are seen on passage. There could well be other unrecorded instances of breeding, because in the northernmost part of the country, the Tell area is dotted with suitable woodland consisting of Holm, Cork and Xen Oaks.

Portugal

The Portuguese region of Tras-os-Montes, the central province of Biera Baixa and the southern provinces of Alentejo and Algarve have the largest oriole populations. The bird has spread into gardens, orchards and various kind of woodland, often in cemeteries. Olive groves, large and small municipal parks, tree-lined avenues and shelter belts alongside agricultural fields, both grass and arable, are also selected at times. In some drier areas, it can be found in Almond orchards, and has been recorded nesting in isolated tall eucalyptus (H. J. da Silva Costa, *in litt.*). Even bushes in steppe-type country can hold pairs (Johansen 1944). In Portugal, tree species used in such habitats certainly include oaks, but records have been made of the use of fig and ash (Bonham & Bonham 1971). Umbrella-shaped Stone Pine also contain oriole nests in sandy parts of the country (B. Martin, pers. comm.). Both in Portugal and in Spain the birds are known to frequent Cork Oak plantations, now a threatened habitat as wineries move to using plastic stoppers.

SPAIN

In neighbouring Spain through to southern France, usable trees include Cork Oak, Spanish Oak, which is taller, and eucalyptus, which is very common all around the Mediterranean. The Extremadura region has a great deal of suitable habitat. Orioles do not breed, however, along the north-west coastal region, and are also largely absent from the Castilla La Mancha region.

Breeding occurs to an altitude of 1,000m in the southern Pyrenees. Breeding in Black Poplar was noted by Pascal Etienne (pers. comm.), who has described a nest in that species of tree high above a deeply banked small river. In the EBCC Atlas, the Spanish population is estimated at 10,000–20,000 (Hagemeijer & Blair 1997), probably making Spain the third most populated European country. Only a small area on the north-west coast and central south-east appears less densely occupied. The small, mountainous country of Andorra, which lies between Spain and France, can boast up to two breeding pairs but a shortage of suitable habitat precludes any more. The Balearic Islands of Minorca and Ibiza support a few breeders, but strangely orioles are absent from Majorca, except on spring passage.

FRANCE

In France in the Drouais region at Eur-et-Loire, orioles nest in alder in the river valleys, and in small woods in dry tablelands, where they show a preference for poplar and oak trees 4.5–8m in height (Labitte 1951). Nesting in Hornbeam has been recorded along the River Loire itself; in 2005, M. J. Read (pers. comm.) described them as 'abundant'. One observer in the Savoie (Alpes) region confirmed that most nests were in oak, but he also found six nests in ash and occasional examples in beech. Sweet Chestnut was also in use there. Apple is often used, especially in orchards. In his northern French Picardy homeland region along the Somme valley, found that, of 70 nests, most were in oak, followed by alder, pine and birch (Etienne, 1988). At one stage he held the opinion that poplars in block plantations were little used, although isolated poplars in mixed woodland with birches and alder were sometimes chosen by the birds. More recently, with the

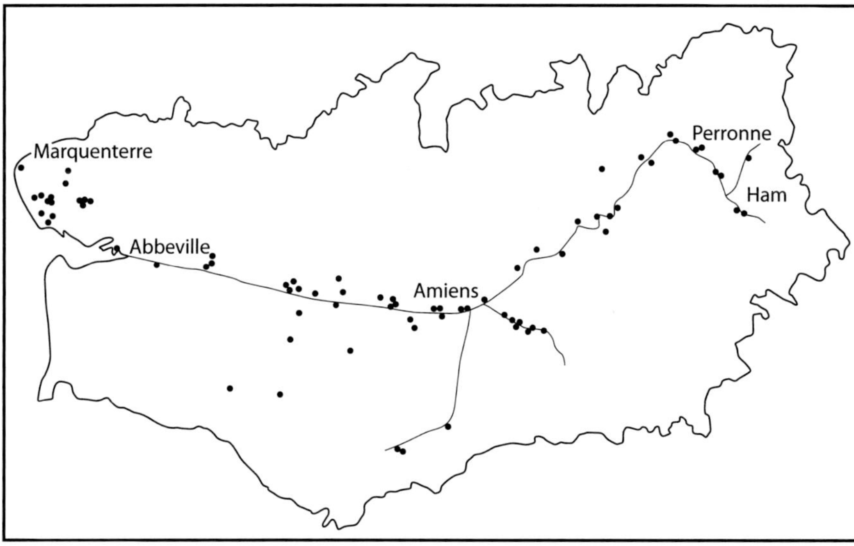

Figure 9. Distribution of oriole nest sites in the River Somme valley. After Etienne, 1988.

increase in the planting of poplars by French landowners, more nests are now being found in this habitat. The river valleys of the Somme, Authie and Canche are now planted extensively with commercial poplar, and orioles have moved into them. Etienne plotted 74 breeding sites along the length of the Somme and its nearby tributaries, producing the map reproduced in Figure 9.

There is an historical reference to a pair having built a nest in a thin beech in the aftermath of First World War battles in that area (Congreve 1918). Considering that there are many large beech forests in this area, it is strange that this is the only record.

During a visit to the Picardy–Somme region in 2005–2006, we noted many commercial poplar plantations, all with orioles present. The river valleys here must be a haven for all sorts of wildlife, with the orioles using the varied habitat to good effect. It was obvious that there were also new plantings, and as the French are noted for good commercial forestry management, the staged planting and felling programmes they adopt augur well for orioles. Some enlightened woodland owners, concerned for the bird's future, are allowing their trees to grow, an indication that the agricultural community are adopting a more environmentally friendly approach. Etienne (1988) recorded that, after the birds had fledged, the family parties leave their arboreal habitat and forage extensively in surrounding marais, coastal marshy areas, such as the one between Ponthoiles and Sailly Bray near Abbeville, where birds were noted searching out food items like insect larvae in reed and sedge. At the seaward end of the river valley at Marquenterre, orioles breed in a mixed birch, alder and conifer woodland, and often forage in the sand dunes, where they collect food not only on the ground but also by picking caterpillars off Sea Buckthorn. Etienne also records that in the dense 4,000ha Forêt de Crecy, several pairs breed in oak near to glades or other clearings. Birds were also noted breeding in the Meuse river valley of Lorraine, where the habitat is mainly hay meadows surrounded by copses and tree-lined borders (van den Berg & Lafontaine 1995). Between Arras and Denain, at the south-eastern border of the Pas de Calais region, peat cutting left a series of ponds where a mixture of oak, willow and alder are now growing, of which orioles have taken advantage. These ponds are connected to the rivers Sensee and Hirondelle and ultimately the Aubigny marsh. Orioles are absent as breeders from the Brittany peninsula, from the Cherbourg peninsula and from most of the Massif Central and French Alps regions (Yeatman-Bertholet & Jarry 1994). As on the Spanish side of the Pyrenees, breeding occurs to a height of about 1,000m. There are other areas of France that are devoid of orioles, such as the open agricultural fields of northern France and the spacious vineyards further south. Nor do the large beech forests seem to attract the birds; this could well be because of the closed canopy as compared with a more open poplar canopy. Further south in the Midi, MR found them nesting in sessile oak

In south-west France, Stephen Marsey (pers. comm.) reports a breeding record from near Carcassone in 2003, where a pair, atypically for France, were seen breeding in a small rectangular plantation consisting of about 12–15 False Acacia. This was set next to a vineyard, which in turn was surrounded by scattered trees, most of which were oak. The birds only foraged in the oaks and returned to the plantation with food.

In south-east France, in the Provence–Cote d'Azur region, an atlas of breeding birds has been produced for 2006, showing both proved and probable breeding in 81 10km squares out of a possible 226. The patterns here show a wider use of the countryside, not just restricted to river valleys as in the north. Approximately 100 of the 10km squares surveyed are land above 1,000m and include a large proportion of the Alpes Maritimes, where there were only a few sites by rivers.

The population of Golden Orioles in France was estimated in the 1970s at between 10,000 and 100,000 (Yeatman 1976), an estimate that had not substantially changed by 1989 (Yeatman-Bertholet & Jarry 1994) However, by 2001 Etienne (pers. comm.) believed they were increasing, certainly in the north of the country. The latest figure is now 150,000–600,000 (Birfield & van Bommel 2006), with a proviso that there seems to have been a decline in the species' fortunes. So the overall picture is a little unclear in France.

Belgium

In 1933, Le Major Georlette (1939) found up to 30 nests in the Namur region of south-west Belgium, all of them in oak, from 3 to 8m above the ground. What was described as an abundant population inhabits an area on the North Sea coast at Zwin. This is an area of an impenetrable mixture of bushes, pines and ponds (van den Berg & Lafontaine 1995). However, Paulussen (1993) states that whereas up to and during the 1960s, in his Turnhout region of north Belgium, each local village could sport tens of breeding pairs, by the early 1970s there had been a significant decrease, and he gives an example of his own village of Zevendonk, where in 1954 he had recorded 60 breeding pairs but by 1991 he could find only seven. By 2005 the position was even more precarious. Wouter Favets (*in litt.*) states that many old and decrepit poplar plantations are being felled and either being replanted with more traditional species or not being replanted at all. In some areas near rivers, this clearance is being done to encourage more marshy areas with a view to re-establishing populations of Reed Buntings and other species. The Belgian oriole population has been variably assessed at 5,000 in 1972 (Lippens & Wille 1972) and 2,300 by 1977 (Devillers *et al.* 1988). Birfield & van Bommel (2006) estimate a figure of 1,600–3,800.

Luxembourg

Neighbouring Luxembourg holds approximately 160 pairs, mostly in the centre and south of the country (Patric Lorge, *in litt.*).

Holland

In the polders at Flevoland near to the North Sea coast, poplar forests were planted by the state, and orioles quickly moved in. Here Wassmann (2004) estimated that 173 breeding pairs occupied 1,446ha, nearly all based on the hybrid poplars. This represents a breeding density of 1.2 breeding pairs per 10ha. Orioles also breed in willow in river valleys, at a density of 1.9 per 10ha in the case of 23 pairs recorded. Research similar to that carried out by the Golden Oriole Group found that not all hybrid poplar cultivars were used by the birds (Bijlsma 1995). We are of the opinion that our fenland population is sustained by recruitment from this Dutch population, as the layout of the poplar plantations on damp flat fen soil is similar. Golden Orioles also inhabit the deciduous woodlands on dry soil in Veluwe. A fen peat colony can also be found around Kemperisland.

Eric Hosking, in his book, *An Eye for a Bird* (Hosking 1970), relates how he erected a small pylon hide at an oriole nest, and continues:

> Our work attracted much interest: even the local Burgomaster and his officials came to inspect our operations and requested me to send them some of the photographs. Probably because of the large number of people who walked through the park, the golden orioles were remarkably tame, thus enabling me to secure some lovely photographs. Prints were sent to the Burgomaster, who had one framed and hung in his parlour.

Year	*Estimate*	*Authority*
1968–73	5,500	Bouman 1974
1977	4,000–5,000	Teixera 1979
1979–85	7,000–10,000	SOVON 1987
2001	*ca.* 4,500	Bijlsma and Hustings 2001
2004	4,000–5,000	Birfield, I. & van Bommel, F. 2006

Table 7. Population estimates for The Netherlands, 1968–2004.

In 2005, however, the position of the poplar plantations has changed. Being mostly hybrids, many of the tress are coming to the end of their lives and are degrading. One large plantation was brought down by very high winds; this was put down to a genetic weakness of the particular cultivar, which had allowed Canker to build up. As the poplars were not planted for commercial purposes, there is no incentive to replace them. Instead, the area has been replanted with a generic mix of native trees. Even if this mix contains tree species that the orioles will use, their growth rate will be slower than hybrid poplars, so any return of breeding orioles will take some time.

As to oriole breeding numbers, in the first decade of the 20th century, the bird was regarded as common but by the 1940s, it was considered a relatively rare species. More recent figures are shown in Table 7.

Germany

North-western Germany has a mixed deciduous-based habitat, where in five plantations covering 83.6ha the breeding density was calculated at 1.9 per 10ha (Flade 1994). Further south in Mittelfranken, out of a total of 60 nests surveyed, 53 were in pine. Conversely in Oberbayern out of eight nests only one was in pine, even though it was the commonest species available. Also in Germany, a nest was found in Cherry (Zippelius 1972). Sabine Baumann (pers. comm.), who has worked extensively on orioles in Upper Saxony as well as in Africa, believes that a decline in numbers in her German study areas is due to fragmentation of habitat, leading to the loss of wildlife corridors. In north Rhine-Westphalia, Wassmann (2004) found that the average territory size was about 0.5 pairs per 10ha, but the birds will forage over an area of possibly 50ha. He also estimates that the population in that province in 1995 was 500–1,000 pairs, breeding in oak, poplar and alder, at nest heights ranging from 3 to 20m.

In the former German Democratic Republic (GDR), Feige (1986) found that, in riverine woodlands, birch was commonly used when near to larger open stretches of water. Near Leipzig, birds were found breeding close by opencast coalmines in poplar trees which had been planted to mask the operations (Beer 1961–4).

Farming in the former GDR was much less efficient than in the west, and this allowed many species of birds to breed commonly. Since the re-unification and the Common Agricultural Policy (CAP) regime taking over, this situation is likely to change, with adverse effects on the wildlife. Fortunately, most of the oriole habitat is unlikely to be affected, as it is mostly willow and birch woodland around extensive lake systems.

Germany's population figures are somewhat confused. Rheinwald (1982) gave an estimate of 19,000–110,000 pairs for West Germany alone. No figures are available for pre-unification GDR, but Cramp & Simmons (1993) give an estimate of 18,500, based on Feige (1986). The total for the reunified Germany is given as 110,000 in the EBCC Atlas (1996). However Wassmann, who composed the text for the Golden Oriole pages in the EBCC Atlas, asserts that his research produced an estimate of 38,000–60,000 breeding pairs for the whole nation. He disputes the accuracy of the methods used previously. The question arises whether the original figures were an overestimate, or if the species has declined since. By 2003 there had been a noticeable decline in north-west Germany according to Sabine Baumann (pers. comm.), but the reasons here are unclear. Birfield & van Bommel (2006) suggest a population of 40,000–90,000 for the whole country and noted a decline between 1995 and 1999.

Austria

In Austria, there are large poplar plantations near Lake Neusiedl in the south against the Hungarian border. The whole area is flat relative to the western half of the country, famous for its mountainous ski resorts. The only remaining large riverine forest is in the Tullnerfeld area on a free-flowing stretch of the Danube. By the river willow and poplar are the dominant species, while in other areas the forest is dominated by ash, oak and elm. Commercial poplar plantations have replaced the natural forest in many places. The Tullnerfeld forest is interspersed with oxbow lakes fringed by reed beds. Another important area encompasses the flood-plain forests of the Danube between Krems and Klosterneuburg. Only small remnants of the original forest communities still exist. Today, poplar plantations and stands of ash and alder cover most of the area. Within this forest, too, there are numerous oxbow lakes and reed beds, as well as hay meadows, arable fields and gravel pits. Lake Neusiedl itself is Europe's fourth largest, and its only steppe lake. It holds some stunning birdlife, orioles are among the commonest breeding birds.

In the swamps near the Danube river, poplar plantations hold a density of 1.9–3.6 pairs per 10ha (Dvorak and Berg 1993). In his dissertation on orioles in mixed woodland at Strasshof near to Ganserdorf, close to the Slovak Republic border, Hovorka (1991) found that Sessile, Downy and Pendunc学ulate Oaks alongside *Robinia* formed the most favoured habitat used, confirming the diverse range of habitats in Austria available to orioles.

Switzerland

In neighbouring Switzerland, Buxton (1932), during the time he was a resident in Geneva, watched birds nesting in the gardens there. He notes that the birds arrive at the end of April before oak foliage is fully open. Then he describes nests in spindly oak about 8–9m high, and also in ash. According to the Swiss atlas for the years 1972–76, most breeding takes place in the north-west Lakeland cantons up to 600m, though in 1975 breeding occurred in the upper Valtellina area at 1,160m. Breeding has also been confirmed along the Rhine near Basle and along the Rhone in Haut-Valais (Winkler 1999).

Italy

The Golden Oriole is a relatively common migrant species in peninsular Italy, where it also nests regularly. Good numbers of orioles breed up to 1,000m in the central Apennine mountain range. They do not breed either along the heel or in a large part of the toe of Italy, apart from a few scattered records from Calabria. Observations suggest that it is less common in the north and east of Italy than in the centre or the west. According to a recently published breeding atlas map (Pazzuconi 1997) they are found in a variety of habitats, and are said to frequent areas where grapes and olives are grown. However, it does not appear that the birds actually use those species to nest in. Fortunately, orioles do no damage to these crops as the fruit only ripens after the birds depart.

In Tuscany, Marco Dragonetti has made a study of orioles in the many oaks around his farm near Magliano in Grosseto province, about 7–8km from the coast. He notes that orioles are more common near the coast than in the interior and can be found in mixed woods. The wood near his house is about 60% oak woodland, mostly Downy Oak, the rest made up of ash, maple and buckthorn. This is a typical mixed wood of Grosseto province. Near the coast, Downy Oak often gives way to holm oak, and the wood becomes maquis. However, Dragonetti's observations suggest that all woods where you can find orioles are adjacent to farmland: vineyards, olive groves, cereal crops and grazed meadows. Hedgerows are abundant, consisting of scattered oaks, with pistachio, Judas Tree, Common Myrtle, brambles, Rosemary and the prickly Christ's Thorn.

Where poplars occur, they are usually small copses along watercourses, and are used by breeding orioles. Other oriole territories found by Dragonetti in Grosseto province include three or four males singing in a mixed wood of willow, oak and cypress near an artificial lake, and another in a pine wood between the coast and cultivated lands. Orioles are more plentiful on the western side of the peninsula than the east. This is because the west is more pastoral while the east is more intensively farmed with open, treeless countryside.

The atlas of breeding birds for the Bologna region gives a rough estimate of 700–1,500 breeding pairs. 40% of these were breeding in riverine localities

Tree species	%
Poplar	25
Oak spp.	18
Walnut	14
Robinia	9
Elm spp.*	9
Willow spp.	9
Hazel	1
Maple, alder, plum	15

Table 8. Tree usage by 78 breeding pairs in north-central Italy. *There are two species of elm in Italy: wych elm (*Ulmus glabra*) and smooth-leaved elm (*Ulmus carpinifolia*). As the structure of *U. glabra* branching looks unsuitable for nesting, it is probably *U. carpinifolia* that is used. From Pazzuconi, 1997.

along the lower, flatter land, the remaining 60% in hilly districts where rivers such as the Reno run. A special point is made that the birds could not be found in beech woods or coniferous plantations and only on rare occasions in chestnut. Only 10% of the birds in this habitat were breeding above 800m and none above 1,000m.

Sicily sports an interesting population in two areas, one of which is on the slopes of Mount Etna to a height of 1,500m (in most of Europe, orioles are rarely found above 800m). Here in the Madonie mountains in the south-east of the island, the birds use copses and mature woods in valley bottoms. In the Iblei mountains nests have been found in Plane trees (Iapichino & Massa, 1989). Sardinia does not appear to have many, if any, breeding orioles but does see quite a strong spring passage of birds (Mike Read, pers. comm.). A similar situation occurs on the French island of Corsica to the north.

The EBCC Atlas (1996) gives a population of 3,000–4,000 for the whole of Italy, but Birfield & van Bommel (2006) estimate that the population is 10,000–20,000. One wonders what the population size would be if the birds were not trapped for gourmet food in the estimated three million roccolo and other traps set. We look forward to a better future, though, as there is evidence that the education programmes of LIPU (Italy's equivalent of the RSPB) are bearing fruit, especially among younger people. Italy has hitherto been known as a country where many passerines are hunted in great numbers for the table. Thanks to LIPU, the general public is increasingly persuaded that such slaughter of birds is totally unacceptable. Although hunting has declined considerably, the hunters, deprived of quarry at home, are mounting hunting expeditions to countries such as Egypt, Tunisia and Bulgaria to pursue their contemptible sport. In 2006, a party of Italian hunters was arrested at Munich Airport (Germany) with 2,643 dead birds (fortunately no orioles) as they were returning from Romania. The birds were destined for Italian restaurants, so the danger is still there. CABS (the German Committee Against Bird

Years	*Pairs*	*Source*
1972	100	Pedersen 1972
1976	200–400	Dybbro 1976
1998	212	Grell 1998
1999	13–19	Breeding Atlas, DOFT
2004	4	Fahlberg 2004
2007	1–2	Fahlberg, *in litt*

Table 9. Population estimates for Demark.

Slaughter) believes that 50 tonnes of dead birds are smuggled annually from the Balkans into northern Italy, and these may well include orioles.

Denmark

Denmark's birds nest mostly on the eastern side of the mainland, and are also found on the islands of Fyn, Sjaelland and Lolland, in what are described as 'well-developed oak or beech woodlands', where the birds nest in ash copses within the stands of beech (Grell 1998). All these sites are coastal. The Danish atlas also shows a few in the south-west near to the North Sea coast and a heavy concentration close to the German border in Sonderjylland and its inshore island of Als. The fortunes of orioles in Denmark are well illustrated by the estimates in Table 9.

The 2007 birds were on Æbelø north of Fyn. This worrying decline in the number of breeding pairs is reflected in declines recorded all along the north-west extremes of the oriole's range. According to recent DATSY Rare Bird Reports, the reason for this decline is a change to more Atlantic weather conditions in mainland Denmark. For example, in June especially there has been 26% less sunshine and 16% more rainfall over a period of 30 years, and the same goes for the region over the German border in Schleswig-Holstein (Fahlberg, *in litt.*).

Norway

There have been a few breeding records from Norway, the first being near Vestfold in 1972. Orioles have also probably bred at the Jomfruland, on the southern coast, described as 'a beautiful and varied island'. The Øitangen Landscape Conservation Area consists of agricultural landscape and oak/hazel forests. Natural pastures, coastal meadows, older broad-leaved deciduous forest, dry coastal scrub and sandy beaches cover the area. A number of IUCN Red List species of plants, insects and birds occur there. Interestingly, the Øitangen Reserve uses a Golden Oriole in its logo, most unusual for a place so far north (58° N).

SWEDEN

In Sweden, breeding orioles were originally considered an extension of the Danish population because of the close proximity of these birds across the Øresund straits. But this may not be true, as the Swedish climate is more continental and can be very hot in summer, whereas Denmark's is maritime, influenced more by the North Sea and ultimately the Atlantic Ocean. Most of Sweden's orioles are in two places; around the two lakes of Vombajon and Krankesjon 20km east of Lund, where the habitat is mainly pine and birch, and 25km north of Simrisham, where beech is the principal tree species. Dan Hammerlund (*in litt.*), who has been monitoring the present population, thinks the Swedish birds are more likely to derive from Polish sources. During the 30 years to 2004, the population has remained quite stable at 20–50 pairs, and in the most recent decade, an average of 70 singing males have been heard. In the past numbers were put at 10 pairs in 1965 (Ernholdt 1966) and at least 50 by 1977 (Sylven 1978), with a possible rise thereafter.

THE BALTIC STATES AND FINLAND

Common throughout **Estonia**, the oriole is also found at a lesser density in the western islands of Kbluma and Sarema. Most Estonian orioles breed near to the Russian border, thinning out towards the Baltic Sea. The birds breed in a variety of habitats (see Table 10).

In 1983–98, large increases (>50%) were observed in breeding woodland bird species, but in the same period, distressing (>50%) decreases were observed in Golden Orioles (Kuresoo & Ader 2000). In fact, Estonia experiences annual fluctuations in oriole numbers, and since 1976 a gradual decline has been observed. However, it may be that since the country became independent of the USSR, habitats have improved by default, ironically because agriculture has declined and because trees have now grown again, especially broadleaved. Ellermaa (pers. comm.) further observes that without EU laws the environment would not be taken seriously in many new member countries like his. Breeding estimates stand at 15,000–20,000 pairs for 1991–2000.

Habitat	*Pairs/ km2*	*Source*
Birch and alder forest	9	Leibak *et al.* 1994
Broadleaved nemoral forest	2–3	Leibak *et al.* 1994
Pine forest	5	Vilbaste 1990
Spruce woods	1–4	Vilbaste 1990
Mixed woodland	3–6	Margus Ellermaa, pers. comm.

Table 10. Habitat use by orioles in Estonia.

Latvian habitat is similar to that of Estonia, and orioles are widespread but not particularly numerous. The atlas map in Visknes (1989) shows greater densities around the Gulf of Riga and to the west of the capital. However, things have changed since the country became independent, as new farming practices have been introduced. Although farming intensity is not even close to that in EU countries yet, it is expected that many private farmers will start, or have started, to adopt western farming practices, which have been a principal cause of the decline of many farmland birds in Western Europe. Golden Orioles have already shown a significant decline in three closely monitored areas, with only a moderate increase in one other (Aunins 2003). The EBCC atlas puts the Latvian population at up to 40,000 pairs.

In **Lithuania**, alder and birch forests are found in the boggy countryside that occurs from northern Poland right through to Estonia. According to Karlavicius (2004), the oriole was widespread in times past and could be quite dense in some parts, but things are changing fast and not for the better. The more recent annual population indices of such long-distance migrants as Common Cuckoo, Icterine Warbler, Common Whitethroat, Wood Warbler and Golden Oriole showed a marked decline in Lithuania during the period 1993–95. Trends of these species in Lithuania correspond with those from Latvia in the same period. Such changes may in part be the result of changes in climatic conditions in these species' wintering areas. Whether this applies to orioles is far from clear, as the declines only seem to be occurring in these Baltic countries and in north-western Europe; the reasons may not be the same in all cases. The two other countries in the region, Sweden and Poland, have not seen declines in orioles.

Finland's sparse oriole population is found up to 65° N, whereas in the rest of its range the Golden Oriole reaches only 60° N. The birds can be seen over most of southern Finland on migration, but the south-east of Finland near the Russian Karelia region is where most breeding occurs. Here there are many lakes, and it is mainly around these that the birds are found. In Karelia, the white birch is the predominant tree in continuous open boggy forests reaching to the Russian border, so these are favourites for use, to a density of one to five pairs/km^2. The bird is also said to use open mature pine forests near to Lake Saimaa in central Finland. There are some occupied 10km squares (according to the Finnish atlas), along the south coast, mostly east of Helsinki. The species was claimed to be numerous in the 19th century but declined disastrously at the turn of the 20th. From the 1940s it climbed back to 5,000 pairs, but, as in Estonia, the numbers vary from one year to another. In 2003, the latest estimate was 3,000 pairs (Tom Lindroos, pers. comm.). Lindroos estimates that there has been a 40% decline over the last 15 years.

POLAND

Poland is well known for its vast wetland systems, especially in the north and east of the country. Here large tracts of marsh and peat-lined bogs are bordered with trees suitable for orioles. The primeval forest of Bialowieza in particular, thanks to its vast area and variety of native trees, is much to the liking of orioles, as well as to many other exciting and sometimes rare species. Orioles are most numerous in park-like landscapes, river valleys and around lakes, especially in aspen woodland, which is widespread in the north-west of Poland. Shelter belts in farmland hold good numbers, with breeding densities ranging from 0.3 to 0.7 pairs/km^2. Notably, along the Vistula river in suburban Warsaw, they are said to breed at a remarkable density of 14–15 pairs/km^2 (Stawarczk, pers. comm.). The average height of nests is given as 6.5m, with a range of 2–20m.

In the Bialowieza forest, orioles breed not only along the edges but up to 7km inside. They occur in mixed groves of oak, lime, hornbeam, ash and alder of various ages, but the nest tree species here have not been recorded. Breeding orioles are also to be found in mixed conifer stands, provided there is birch as well, which may indicate that it is the latter in which they choose to nest. The forest researchers measure the breeding rate census plots of 25ha, and usually 0.5 pairs per plot are counted, but in some sections four pairs per plot have been registered. There seems to have been no change in the population in the forest, at least from 1975 to 1999. This estimate is not as certain as in the case of typically territorial species, however. A radio-tracking exercise would be necessary to check these results (Tomialojc, *in litt.*).

Another valuable area of primeval forest is under threat at Augustow, where a new motorway is planned. Whether this would affect orioles there remains to be seen, but there are other birch groves nearby that are used, and the motorway itself will create margins in the trees that may even be to the liking of the orioles.

Elsewhere in south and west Poland, the Odra river valley is lined with woodland near Wroclaw (Gorman 1994). Here oak, beech, alder and willow with open areas and backwaters hold excellent numbers of orioles and near to the Ukrainian border the Chelm Marshes, where grazing meadows adjoin the peat bog marshland, hold many bird species, including good numbers of orioles. In fact, orioles breed over the whole country, but not above 500m. Wassmann (2004) assesses the Polish population at 15,000–30,000 pairs.

Czech Republic

Golden Orioles are recorded in every part of the Czech Republic to varying degrees. The Czech Republic has remnant riverine woodlands and large areas of fishing lakes in Bohemia and the southern half of Moravia, along the Morava and Dyje river complex in the east – habitat suitable for orioles. In the west, the Sumava forests and peat bog, areas that lie along the border between Germany and the former Czechoslovakia, were left to run relatively wild, so that all sorts of birdlife took advantage.

The Trebonsko region is characterised by flat ground with stepped edges, mainly in the southern part of the area. In 1979 part of the region, an area of 70,000ha through which the Luznice River flows, was designated a protected landscape area,

Area	*Pairs/10ha*	*Habitat*
Central Moravia	3–4	deciduous, coniferous and floodplain**
Southern Bohemia	0.3–5	300–400-year-old oak woodlands
South-west Bohemia	0.3–2.5	'palace parks', e.g. Livejovice
Northern Bohemia	1.4–1.9	abandoned orchards, riverine stands
Northern Moravia	1	floodplain forests
Northern Bohemia	0.6	'palace park' at Krasny

Table 11. Variation of breeding density with habitat in the Czech Republic. **Central Moravia is rich in variety, the three habitats occurring in different parts of the region. The deciduous forests are of oak, birch and *Robinia*; the coniferous of pine and spruce (admixed with oak).

and in 1977 was given the status of UNESCO Biosphere Reserve. The creation of a network of large lakes and carp ponds, some dating back to the 16th century, changed this formerly monotonous moorland to an area of great variety, which is unique in the Czech Republic. Also, many centuries ago, oak, mainly sessile and downy, was planted along the banks of these sometimes quite large lakes, and orioles have taken full advantage. The woodland is well spaced with many glades. Some of the lakes are edged by oak colonnades, and to walk here and listen to oriole song while watching Collared Flycatchers and Middle Spotted Woodpeckers is a memorable experience.

Breeding densities vary with habitat, as Table 11 illustrates. As for tree species used, out of 39 nests recorded, the findings were as follows (K. Stastny *in litt.*);

Species	Nests
Oak	6
Robinia	5
Alder	4
Ash	4
Others	20

Height of nests ranged from 2–18m, distributed as follows (K. Stastny *in litt.*);

Height	Nests
2–3	6
3–4	4
4–5	6
5–10	5
18	1

Slovakia

In Slovakia the extensive flood plains to be found along the River Danube contain rich habitat for Golden Orioles. As elsewhere the best places are the riverside woodlands. North of the capital, Bratislava, is the Carpathian mountain range. On its lower southern slopes, which are covered in beech forests, the clearings, orchards and vineyards are suitable for orioles. On the border with Hungary lies the state nature reserve of Zadielska Dolina, near to which Golden Orioles are regarded as common. The reserve contains a limestone gorge with small river valleys, oak woodland and meadows. In the east of the country, there are more flood plains. Although they have largely been drained and converted to agriculture, there are riverine valleys and lakeland edge habitat that allows orioles to breed in reasonable numbers. The city of Kosice also has orioles nesting within its confines and a City Environment Report included detail on the birds nesting in *Robinia*, in other tree species like poplar and willow in a waterlogged area, as well as in a poplar screen planted around a metallurgical factory. According to Cramp & Simmons (1993) the population for the whole of the old Czechoslovakia is 'stable'.

Hungary

Breeding orioles are common throughout Hungary in a variety of habitats. Close to the Austrian border there is a shallow saline lake, Lake Ferto, which is approached through a damp thicket area. Here orioles accompany other species, such as Common Nightingale and River Warbler, that enjoy such habitat. Another salt lake, Lake Peteri, south of Budapest, is a similar type of habitat that is to the orioles' liking. In northern Hungary stands the Borzsony Hill, with vineyards on the lower slopes, and allotments and open gardens all adjoining human habitation used as summer holiday houses. The vineyard owners may regard orioles as a pest if they gorge on too many grapes, but this is not likely to happen until the birds prepare to migrate south in autumn.

Some 20km south-east of Budapest, near a village called Dabas, there are poplar plantations. Here our birds are neighbours to Black Woodpeckers, amongst others. On a peninsula on the northern banks of Lake Balaton, itself a holiday resort area, stands the village of Tihany. To the west of here there are more wooded areas where orioles announce their presence with song. In north-east Hungary, there are flood plains along the upper reaches of the Tisza River. Regrettably, primeval forest habitat has gradually been replaced with commercial poplar plantations; regrettable, that is, for much of the wildlife that would have been there before this process, but this of course has benefited the orioles, which are there in good numbers. Breeding orioles are also regular in public parks, where they have been recorded nesting in false acacia, amongst other parkland tree species (Gorman 1994).

EBCC figures for Hungary stand at 100,000, but Wassmann (2004) reduces this figure to 40,000–70,000. Latest figures from a 1999–2000 survey indicate a drop of 12% between the two years in the areas surveyed (Szep & Nagy 2002).

Romania

Romania, lying as it does between Hungary and Bulgaria, two countries with big oriole populations, is another eminently suitable place for them. However, it is a more mountainous country, much of it alpine or subalpine in nature. Orioles do breed, however, up to about 1000m above sea level in the foothills of the Carpathian Mountains. The most heavily populated parts are near to the Black Sea and in the Danube delta region. The wooded Babadağ hills are said to hold several pairs, and a reserve at Hagieni in the south-east is another good place as it contains oak woodland, mainly *Quercus robur* and *Q. pendunculifolia*, with wet meadows and a small river valley. An area of orchards and vineyards near to Lake Oltina further inland and, on the western border, woods around the fishponds at Cefa provide similar habitat to that in neighbouring Hungary. With such a dense breeding population nesting largely in wild poplars – both *nigra* and *alba*, plus many in hybrids such as robusta, deltoid, romboid and picea – many birds have to make use of less-favoured species such as apple, plum, mulberry and lime. As in Hungary and Bulgaria, a considerable amount of false acacia is used, a tree that was originally planted in gardens and parks but has now gone almost rampantly wild. Mulberry is very common in south and central Dobruja in southeast Romania, and here it is estimated that 30% of oriole nests are to be found in mulberry. This is the only central European country where lime has been recorded being used. Attila Sandor (*in litt.*) says he has found oriole nests in lime, ash, false acacia, black and white poplar, oak, alder, plum and mulberry.

The differences in overall habitat in Romania may be the cause of widely differing population estimates: 80,000–90,000 according to the EBCC Atlas, but a much larger figure of 680,000–890,000 is given for 2000–2003. According to Sandor, this may simply reflect annual fluctuations. He notes, for example, that 2003 was better than 2004. *Birds in Europe* (Birfield & van Bommel 2006) confirms the latter figure, with the trend being stable.

Bulgaria

Bulgaria was long thought to be the country most heavily populated with orioles, and certainly their presence is very obvious up to 700m, possibly even to 1,500m. The preferred tree species used by orioles in Bulgaria is poplar, much of it commercial, with *Robinia* being second choice. In third place is a mixture of ash and secondary oak. Suitable habitats are found along rivers and other wetlands, in parks within towns and villages, and in open woodland. Habitat conditions have not changed greatly for a long time and agriculture is not very intensive, both of which factors are of benefit to wildlife (P. Tankov, *in litt.*). Some birders visiting a hotel in the country reported that it was the first bird many of them saw on arrival there, and as they watched from their bedroom windows, the bright yellow males were much in evidence; the early morning dawn chorus is dominated by oriole song, and so loud as to wake visitors (B. Martin, pers. comm.).

Good places for orioles include the Ropotamo valley, a lush habitat with suitable broadleaved trees. Holiday resorts where many birds are reported nesting in built-up areas include the large complex at Borgas and the Golden Sands resort north of Varna, again on the sea coast. The fear is that tourist developments are increasing, to the detriment of the natural environment. The Varna resort, which is on land that juts out into the Black Sea, is an important migration watch point; orioles passing through in August and September are described as in their 'millions' (Lyubo Profirov, *in litt.*).

The bird is widespread throughout the country, mainly up to 700m above sea level, but some may reach 1,400–1,500m. Population estimates are put at 300,000 pairs in EBCC atlas charts. More local sources, on the other hand, put the population somewhat lower at 150,000–200,000 pairs.

Greece

Golden Orioles are mainly confined to northern and central parts of the mainland as a breeding species. In Greece, most breeding orioles are found to the east and west of the northern part of the central mountain range and on the lower fertile lands of Thessalia and Thessaloniki on the Aegean, and between Thesprotia and Preveza on the Adriatic coast. In southern Greece the oriole, known locally as the figbird, is generally a passage bird in autumn. Isolated pairs or small populations exist in pockets of suitable habitat in Ipeiros, Thessaly and Sterea Hellas, but higher densities distributed in a more uniform manner may be seen only in the uplands of western Macedonia and parts of eastern Macedonia and Thrace. There are still no definite breeding records from the Peleponnese, though a few pairs may perhaps nest in the north. Nor is there definite proof of breeding on any of the islands, except Thasos off the Macedonian coast and occasional pairs on Lesvos and Samos. Orioles breed in a variety of woodland or open park-like habitats with scattered trees at altitudes up to 1,200m and prefer mature and open deciduous woodland dominated by oaks. They are also common in patches of riparian forest, in commercial poplar plantations, and very occasionally in coniferous woodland. During the breeding season they take advantage of cherry trees, both cultivated and wild, and mulberries. As Greece is on a major migration route to and from other countries, many records of orioles refer to passage birds (Handrinos & Akriotis 1997). The total breeding population is tentatively estimated at 5,000–30,000 pairs. Cramp & Simmons (1993) simply refer to 'a few thousand pairs' for the Greek population.

THE BALKAN COUNTRIES

At the time of writing, information on **Bosnia and Herzegovina, Montenegro** and **Kosovo** is hard to come by. Much of the terrain is mountainous, with numerous river valleys. These contain varying amounts of riverine woodland suitable for orioles. Meinertzhagen (1923) noted that birds arrived from 23rd April 'in the better wooded parts of Montenegro and Dalmatia'.

In **Serbia** the Golden Oriole is numerous in Belgrade's well-wooded sections in the north of the city, along the bank of the Sava River, where males sing, marking their territory from May to late summer (Dajovic 1998). However, there is much suitable wooded habitat all over the country, along river valleys and in open agricultural areas, which contain tree-lined fields. Serbia figures provided by BirdLife International include those of Montenegro and stand at 100,000–140,000 pairs.

The figures for **Croatia** are given in the EBCC atlas at approx 3,000 pairs, but BirdLife later put them at 50,000–100,000. Orioles are found in woodlands along the Adriatic coast and on offshore islands. In the inner part of Croatia, habitats vary from deciduous to Mediterranean pine forests, but no specific studies have been undertaken there (J. Muzinic *in litt.*)

The oriole population in **Slovenia**, estimated at 4,000–8,000 pairs, is widespread at low concentrations. Orioles are found in at least 66% of the 10km squares and mostly in the lowlands, with a lower density in middle altitudes. At Ljubljana Marsh SPA in the east of the country, a breeding density of 0.8 pairs per 10ha has been calculated (A. Figelj *in litt.*).

In the **Republic of Macedonia**, orioles breed nearly everywhere, being recorded in orchards, oak forests and poplar plantations at up to 1,000m (B. Micevski, *in litt.*)

UKRAINE

Ukraine is known to hold a considerable but decreasing population, according to BirdLife International figures. With regard to studies in particular habitats, Kharkova and Boehme (2005) carried out research in Les na Vorskle Nature Reserve's high oak woods (Bolgogrod region, technically in Russia but only just over the Ukrainian border). They found that there were eight oriole nests in the study site, of which seven were placed in oak, and one in small-leaved lime. Heights were:

Range	*n*
2–4m	1
4–6m	5
6–8m	2

With many trees in this wood reaching heights of up to 16m, the birds were obviously not using the high ones as they sometimes do elsewhere, including in the fenland study area in the UK. In the Kharkov region of the Ukraine, an estimate from 1940 gives six to 14 nests per plot of 25ha, which is a high density (though the habitat was not specified) (Kharkova & Boehme 2005).

Kostin (1983) notes that in the Crimea orioles are a breeding bird of river valleys in the foothills and at some sites in steppe, but in the rest of the territory, including the mountain–forest part, it is only a passage bird. At nesting time it occurs in the plots of trees near river mouths, especially if there are large trees. It is comparatively common in the valleys of Sakgir and Biyuk-Karasu, and is often recorded in the foothill plots of Burulchi, Kuchuk-Karasu and Mokryi Indol. In this region, passage numbers are low, and in some years it is no more than a sporadic breeder. In other words, orioles breed sparingly in southern Ukraine, especially on the Crimean peninsula, but gradually increase to the north as suitable habitat increases, such as the large wooded areas of the Carpathian Mountains. This is where the Carpathian Biosphere Reserve is located, an area covered mainly by forest vegetation, which occupies nearly 90% of its total area.

In the flattest and warmest parts of Transcarpathia more than 25 tree species form the forest cover. The oak forests are mostly sessile oak, but with some *Quercus polycarpa* and *Q. dalachampii*. These would all provide suitable nesting habitat for orioles. After pine at 33%, oak at 24% is the second most common tree species in the Ukraine. The remainder are species that the Golden Oriole has not been recorded as using anywhere before, so it might be safe to assume that oak is also the most commonly used tree for nesting purposes. Pedunculate oak is the characteristic oak species of most upland habitat in Ukraine, especially the Kuziy massif, at an altitude of around 1000m. BirdLife International gives a population estimate of 140,000–190,000 pairs.

Belarus

Because Belarus borders Poland to the west, much of the wetland lacustrine habitat spills over, and it also shares part of the vast Bialowieza primeval forest. As a result, birds are quite numerous. BirdLife International in Minsk regards breeding oriole as 'quite common'. On his visit to the country to advise on wetland projects, Norman Sills of the RSPB (who is also a member of the Golden Oriole Group), found that orioles were very common in the area of the Pripet Marshes, inhabiting villages and gardens, where they appeared to use mainly poplar belts and copses. He also received a report of a pair nesting in a small village called Turov, in a mature willow in a garden that backed on to the bank of the River Pripet. Given the prevalence of fish ponds and similar marshy habitat mixed in with agricultural land containing field shelter belts – all habitat similar to that of neighbouring countries like Slovakia and Poland – the oriole population is probably dense to the point of saturation.

A large part of the Bielowieza National Park is situated in Belarus (1250km^2). There are 20 forest types, four of water plants, two shrub communities and 13 areas of peat bogs and meadows. All the major forest associations found in this part of Europe occur, some being represented by Eastern European forms, such as lime and hornbeam communities, whilst others by central European forms like oak and hornbeam; dominant tree species include spruce, hornbeam, lime, alder, oak, maple, ash, birch and aspen. As in Poland, the birds are likely to use most of these in open situations. There is an absence of beech, yew and larch, but this is of little consequence for orioles. The BirdLife International population estimate is 130,000–160,000 pairs, with the population stable.

CHAPTER FOUR

Habitat in other parts of the breeding range

THE CAUCASUS

Armenia

From observations made by ornithologists on their trips to Armenia (as reported in various OSME publications), it appears that this small, landlocked, mountainous republic is well-populated with orioles, given its size. A visit in 2001 recorded them in poplar trees at Dilijan, at Goris, in the foothills of Mount Garma and all along the Garni gorge to Azat Zabasor. In Yerevan city several were found around the Hrazadan River Gorge. Usually they are found in mature woodland, groves and poplar plantations. Nests are found in poplar, walnut, and, recorded for the first time in this survey, pear, less often in willow, apple and peach (another first). Special note is made of a pair in a small pine stand near Gndevaz power transformer station.

As the elevation of the whole country averages 1500m, suitable habitat is in short supply. A population estimate of 1000–1500 pairs is given, and the general opinion is that it is 'uncommon'. This is probably why interest is taken in the species here, and it is proposed that a more detailed survey be undertaken (AUA 1999).

Georgia

In Georgia, the oriole breeds principally in the agricultural lowlands at the southern and eastern ends of the country, to the Azerbaijan border. It is common around village gardens and in riverine situations, where it uses poplars as well as walnut and pistachio. This is the first time the latter two have been encountered as nest tree species. Drier savannah types of forest have also been recorded in use by the birds (Z. Javakhishvili, pers. comm.).

Azerbaijan

The Golden Oriole is regarded as common in Azerbaijan. The latest population estimate puts them at 50,000–100,000 pairs for 1996–2000. It breeds to about 500m in hills up to the upper forest belt and down to the Caspian Sea coastal towns, including the capital Baku in its parks and gardens. Azerbaijan has more birds than the other two Caucasian countries, perhaps because most of the land is flatter and at a lower elevation. Poplars and various nut trees are again favoured nest tree species. Studies carried out from 1986 to 1990 during May, June and July in Azerbaijan show that the Golden Oriole breeds in light forests among human habitation and gardens (Mukhtarov 1991).

TURKEY

At the time of writing, a breeding bird atlas for Turkey is being researched by the country's equivalent of the RSPB, Doğa Derneği, so data are not complete. However, interim results show that the bird breeds over most of the country, but perhaps to a lesser extent in the central eastern region, where there are more open, steppe-like habitats. The birds use poplar to a great degree around farms and forest plantations. The Turks use poplar a great deal for timber. Turkey has widely varying climatic and topographic conditions, which result in a broad diversity of forest types. European black poplar has a wide distribution in Turkey and has been cultivated by farmers on private lands for centuries. It makes a considerable input to both the rural and the national economy. There are approximately 130,000ha of poplar plantations, of which 70,000ha are hybrid poplars and 60,000ha consist of various clones of black poplar. Black poplar provides about 57% of Turkey's 3.5 million cubic metres of annual poplar wood production. The wood is used for a wide range of purposes, especially furniture, floorboards and plywood. A very large proportion is also used by rural inhabitants for house construction. There are many poplar plantations in central, eastern and south-eastern Turkey, as well as shelter belts around fields and along roadsides. Along many river banks often in belts from one to three rows wide on both banks create ideal habitat for orioles. As in Belgium there is much trial work undertaken to breed new clones suitable for the county's soil and climatic conditions. There are also still at least five sites in the Anatolia region where the original black poplar survives. From this it is clear that the oriole population is secure in poplar alone, but in south-western Turkey, around Köyceğiz, the birds also use oriental sweet gum (G. Welsh, *in litt.*).

This area supports 50% of the remaining population of this tree species, its range having declined by 80% since 1949. Nowadays it occurs in scattered patches. Geoff Welch, who has been assisting with the new atlas work, says:

> I think the main reasons for its attraction to orioles are that it does have large leaves, forms dense stands, usually in wet areas, with open areas between the blocks of trees so there are plenty of nesting and feeding opportunities, and is another of the few trees that are able to grow to a reasonable height.

As the atlas work is on-going, there are no up-to-date figures for the breeding population, but BirdLife International (Birfield & van Bommel 2006) estimates a population in the range 350,000–600,000 and indicates a probable decline.

CYPRUS

In Cyprus, the Golden Oriole is a fairly common spring migrant, sometimes in flocks of 100, as at Cape Andreas on 23 April 1973 and 80 on the same date at Askamas in 1978. It is much less common in autumn. Golden Orioles have been recorded as breeding twice, in 1958 and 1989, and both pairs used oriental plane, similar to that used in Sicily and at an elevation of *c.*1100m. There are records of singing in the breeding period elsewhere, but unfortunately a pair that had been seen mating and displaying, in 1971 near Kyrenia, was shot (Flint & Stewart 1983). It is to be hoped that EU rules will be properly enforced in future, so that orioles (and other species) have a chance of breeding successfully.

THE MIDDLE EAST

Lebanon

Orioles are seen on passage, often in sizeable numbers. Benson (1970) reports that in the Bekaa Valley area when they appear on migration they are shot out 'almost to the last bird' from the tall trees that are only to be found around villages and towns. Otherwise the general countryside is quite bare, so orioles are drawn to the trees, seemingly on a suicide course. They do breed occasionally in the south of Lebanon, using eucalyptus, as in Israel (G. r-Jaradi, *in litt.*)

Israel

In Israel, breeding has occurred almost annually since first recorded in 1983, when two pairs were found in some tall, 100-year-old eucalyptus trees in HeMayssedin Wood near to the Hula nature reserve in northern Israel; the population is now estimated to have risen to five pairs (Y. Perlman, *in litt.*).

Iraq

It is certain that many migrating birds pass through Iraq, but because of the situation there at the time of writing any studies are impossible. However, old records show sightings near oil pipeline pumping stations and others where there were any groups of trees. These were mainly spring migration records. There may be a small enclave of breeders in northern Kurdistan according to old distribution maps, but this cannot be confirmed.

Syria

In Syria, the oriole is entirely a passage migrant, to a greater intensity in spring than in autumn (Serra 2005). Reports of breeding which come from the Ras-el-Basit area in north-west Syria are unacceptable by modern standards, so cannot be confirmed. There is also an instance of birds settling near Damascus in 1982, but they later 'disappeared'. Gianlucca Serra (pers. comm.) still believes that breeding cannot be ruled out.

IRAN

In Iran, orioles breed in similar conditions to those found in the three Caucasian countries, but the birds are more widespread at lower levels down to the central and western parts. They are also regarded as regular breeders in the Golestân province at the south-east end of the Caspian Sea. There is little information about which tree species are used. Orioles breed around Tehran and along the Caspian Sea coastal region from the Azerbaijan border to near the Turkmenistan border, where it is possible they might contact the *kundoo* oriole in small numbers. No population estimates are currently available. It is also possible that a few pairs breed just inside the Kurdistan region of Iraq.

SAUDI ARABIA AND THE GULF STATES

Perhaps surprisingly, Arabia has breeding orioles, but only in small numbers. They are recorded where farms have been established in low-lying areas where it is possible to tap into aquifers in order to irrigate crops and grass for animals such as dairy cows. Trees, mainly eucalyptus, have been planted around these aquifers to provide protection from harsh winds, and the birds have found them to their liking. The main places where this has occurred are near coastal regions of the Eastern Province and near to Tabuk in Kuwait, where orioles have bred in a villa garden, and at Ha-il in the centre of Saudi Arabia, where a nest was found in eucalyptus. At the Abdulla al Habashi farm near Jahra, two 'not fully grown' juveniles were seen on 6 June, and by 12 June an adult female in company with a juvenile were also recorded. Where the nest was placed was not determined, and initially it was thought it might be in an apple tree. However, there were also eucalyptus trees in the vicinity.

The species is thought to have potential to spread because more farms and parklands are being set up. Birds are regularly seen throughout the summer in the Saudi capital Riyadh (Tom Talbot, pers. comm.), but he has noticed that the taller eucalyptus trees are lopped for fear that they might fall in high winds from the desert, probably lessening the likelihood of breeding here. Mike Jennings (pers. comm.) believes that breeding could well increase in the Arabian peninsula because of habitat creation consequent upon the establishment of more farms in otherwise arid situations. More parkland is also being created. He estimates that the number of breeding pairs by 2005 had reached 12, and that the potential for expansion was good.

In Oman, Dubai, Bahrain and Qatar, the species is regularly seen on both spring and autumn migration, probably journeying to and from Iranian breeding sites. They are sometimes also found on islands such as Das Island in the Arabian Gulf. Shooting of birds does, unfortunately, occur: for example, out of a total of 1,417 birds of various species shot during the autumn passage in 1999 at Tuhla in Kuwait, seven were orioles.

THE RUSSIAN FEDERATION

Russia is divided between the continents of Europe and Asia. For the purposes of this book we shall split it in two, using the Ural mountains as the dividing line, which is also the line along which the Palearctic zoogeographical region splits into West and East. East of this line is Siberia, where there are two *Oriolus* species: our European Golden Oriole and, in the far east, the *diffusus* race of the Black-naped Oriole, often referred to as the Chinese Oriole. The Eurasian Golden Oriole inhabits the European part of the Russian Federation and breeds from the south in the Stavropol/Caucasus region to Karelia province in the north. The distance

Figure 10. Map of western Russia, showing places discussed in the text.

between the two ends, north to south, is about 2,500km, so habitat and climate change enormously between the two. It testifies to the adaptability of the Golden Oriole that it can breed across this whole range. In the European part of the Russia, the Golden Oriole breeds as far north as 60°N, but in the taiga zone it is rare everywhere. In the forests of the southern belt, its numbers are considerably higher.

Northern Russia

In northern Russia above the latitude of St Petersburg, Golden Orioles are a comparatively rare species, reaching up to the latitude of Medvezhiegorsk, 63° N. The greatest density of nesting is found in the southern half of the Pudozh district, Obonezhie, Prisvirie and the south-eastern shore of the Ladoga Lake, which is 4° further south. Annually, five to 10 pairs are recorded on Valaam Island. On islets further north in the Ladoga, however, the Golden Oriole has not yet been recorded, while to west of Valaam Island it had been found only once, so this represents the most northerly point at which the orioles breed. Given the relatively short summer daylight at this latitude, they need to get down to breeding immediately on arrival. The temperatures rise considerably, the humidity is high and showery rain is regular.

The preferred habitats are the pasture copses and the forests with small trees on the margins of agricultural areas. Thin stands of almost-mature pine and birch trees and patches of coniferous–deciduous plantations between overgrown, cleared spaces also feature strongly. In Pudozh, orioles were recorded nesting in the coastal park near the Zhuravki River floodplain. Breeding densities are given in Table 12.

The distribution and number of breeding pairs have not substantially changed over the last few decades, though there has been a decrease in western parts of the breeding range, where orioles were comparatively common at the beginning of the century (Artemeyev *et al.* 1993). The oriole is reported to be annual in small numbers in the Lake Ladoga area, both on passage and breeding. The preferred habitat varies but always seems to include mature deciduous trees. There are few records of breeding in pine woods (Noskov *et al.* 1981).

The Golden Oriole is widespread and breeds throughout the entire territory of the St Petersburg region, although it is not abundant anywhere there. In the northern parts, the bird is at the edge of its range and breeds at only a few sites,

District/Town	*Habitat*	*pairs/km²*
Pudozh	meadow-field habitats with forests and floodland	1–2
Pudozh	pasture copses with areas of low bog	2–4
Petrozavodsk	groves of deciduous plantations between fields	2.5
Konchezero	pasture forests, in aspen, bird cherry and alder	4.8

Table 12. Breeding density in the Pudozh district, Russia.

particularly in Priladozhie (Lake Ladoga region), along the Svir' River valley and in Prionezhie (Lake Onega region) (Mal'chevski & Pukinski 1983). In southern districts and in old parks near St Petersburg, where the presence of this bird was always noteworthy, its numbers are low. In those places where regular observations were carried out (the parks of St Peterhof and Pavlovsk and the vicinity of Merevo village, for instance), the number of males singing in June in comparison with the end of the 1960s has dropped by approximately two-thirds. In 1978, in the 600ha of Pavlovsk Park, a total of three pairs was recorded. The reason for the decline is unclear; local opinion is that it is a temporary phenomenon. Outside the borders of the St Petersburg region in southern Karelia it is very rare (Ivanter 1962), but in the Vologda region the oriole is common, and in recent years it has even increased in numbers.

Oriole nesting sites in the St Petersburg region are quite diverse. In varying degrees, the birds nest in deciduous, mixed and coniferous (pine) woods. They show a marked preference for semi-mature and mature plantations, and also old parks. Orioles are also found nesting in forest edges overgrown with young, mixed trees. In the parks of St Petersburg, orioles breed only occasionally. In this region, orioles place their nests in both deciduous trees and pines. Shnitnikov (1949) discovered that, of the deciduous tree species, birch was the most popular. The rest of the nests were found in oak and bird cherry. The height of nests varied from 3.5 to 18m. After the end of July, the birds leave the natal area, and frequently turn up in clearings near fruit-growing fields and on forest edges, overgrown with raspberry.

Central Russia

In central Russia, the Volga/Kama region is centred on the Volga river and its tributary, the Kama. It lies around 56–58° N. As with most of central Russia, it gets showery rain almost every other day, with thunderstorms that pass quickly occurring about once a week during the time orioles would be breeding. Having said that, there are years of little rainfall. The Golden Oriole breeds throughout the whole territory. Golden Oriole numbers are not high in the north of the region and in the taiga zone they are even lower, but to the south, in lighter, open forests, numbers increase.

Area	*Pairs per ha*	*Reference*
Sretenskoe Forest, Mari Region	0.5	Pershakov 1932
Kinel River Valley, Kuibyshev Region	0.7	Mal'chevski. & Pukinski 1983
Nr Nizhny Novgorod	0.9	Shaposhnikov 1938
Timashevski	1.5	Mal'chevski. & Pukinski 1983

Table 13. Breeding densities in central Russia.

Orioles in this region prefer light, open deciduous forests where birch, aspen and oak predominate. In the context of dealing with such a large country one has to understand that all these places may read as though they are near to each other but actually may be hundreds of kilometres apart.

It is sometimes difficult to assess how density figures in the case of orioles are estimated. In large, open woods it could be easy enough, but when they breed in shelter belts, the term 'per hectare' seems irrelevant because the birds will fly from belt to belt to feed, even if enough trees are present, and the area can sometimes be quite large, as we have noted in the fenland study area. If the belts are close to each other, the area will be smaller but still cover a similar number of trees.

A summary of the habitats used in central Russia and their perceived breeding densities is given by Mal'chevski & Pukinski (1983). This is shown in Table 14.

In the field shelter belts of Zavolzhie (Trans-Volga Territory) the Golden Oriole plays an important role and it should be considered as the most 'useful bird of forest belts' (Mal'chevski & Pukinski 1983). Quite often orioles concentrate at breeding sites of certain insects. For example, in Buzukul forest (Orenburg region), in oak woods infested by *Bombyx* larvae, oriole numbers were 20 times higher than in neighbouring forests (Shilova 1953). However, it is known that orioles are one of the few species of birds that willingly nest in the tall tree belts of insignificant width and even small density.

Further south, orioles breed more commonly and are even present in the outer suburbs of Moscow. In the Moscow region as a whole, the distribution is mainly to the east and south, with a smaller concentration in the north. The region was surveyed by a total of 401 observers and professional ornithologists between 1999 and 2004. The total oriole population was put at 109 confirmed breeding pairs and eight probables. In the capital itself, 15 breeding pairs plus one probable were recorded (Kalyakin & Voltzit 2005). In the Zavidovo National Park and the Upper Volga, Nikolaev (1998), in his treatise on the birds of marshy landscapes, says that orioles occur in deciduous plots of marginal and island forests in marshy areas. In the pine plots of peat bogs, the Golden Oriole is very rare. Sometimes it can be seen

Habitat	*Pairs per ha*	*Mean range*
Aspen woods with admixture of other tree species	7.4	2.4–16.6
Birch woods with admixture of other tree species	7.2	3.8–14.3
Oak groves with admixture of other tree species	7.1	1.0–26.6
Overgrown, burnt-out sites; cleared spaces; shrubs	5.3	0.9–11.1
Pine woods with admixture of other tree species	5.2	0.9–14.3
Mixed deciduous forests consisting of other tree species	4.9	1.3–12.5
River valleys, swamped and flooded deciduous forests	3.6	0.4–10.0
Spruce forests with admixture of deciduous species	1.8	1.1–2.5

Table 14. Breeding density and habitat type in central Russia. From Mal'chevski and Pukinski (1983).

Habitat	*pairs/km*^{2}
Overgrown peat fields and peat cuttings	0.5–1.2
Mixed forests	1.8–3.4
Forests in floodlands	2.0–3.0

Table 15. Habitat preferences in northeast Moscow. From Gerasimov *et al.* 2000.

in the pine ridges and groves that contain birch trees, in the pine–birch, lake-edge plantations. Observations of birds in the pine–*Sphagnum* bogs have been recorded along the border with Estonia and the Vologda region of the Russian Federation (Kumari 1952). It is more regular in tree stands with a mixture of birch and black alder on the low marshes, in the pine–birch and alder–birch edges of peat diggings and along drainage channels.

In the Ivanovo region north-east of Moscow, the oriole is also a common breeding species, inhabiting mainly tall deciduous and light mixed forests (Gerasimov *et al.* 2000). Habitat preferences are detailed in Table 15.

Kaliningrad

There is an adjunct province of Russia on the coast of the Baltic Sea named Kaliningrad, which has similar habitat to its two neighbours, Poland and Lithuania. There does not appear to be any published work to hand as to the oriole population there but it is unlikely to be very substantial. Oriole records mostly derive from ringing that is carried out at Rybachy, where there is a migration watchpoint and a promontory called Courish. About 20 orioles are ringed each year, and there have been three retraps, which could point to local breeding.

Southern Russia

In the far south of the country is the North Caucasus region, ranging from Stavropol to the northern slopes of the Caucasian mountains. The Caucasian mountains connect the Black and Caspian Seas. This is largely an agricultural region, with river valleys and stands of trees and shelter belts around fields in which there are many orioles (J. Gordon, pers. comm.). Poplars are very numerous, and oak trees are mainly at higher elevations, where they become more stunted than in places further west. Orioles can be found on the lower slopes of the mountains but probably at a lower density.

In the east and centre of this region lies a place of great biodiversity called the Mynach wetlands. Mynach is an ancient river and lake system which was connected to the Caspian Sea in past times. Mynach also forms the geological and geographical border dividing Europe and Asia. The Mynach is an area of lakes and marshes surrounded by deciduous woodland. Although no population densities have been measured, this is likely to be the region with the highest density of breeding orioles in Russia. In eastern Caucasus, however, there are fewer tall trees, and here Valery Moseykin (pers. comm.), who studied and photographed the birds near a village named Kievka, just south of the Mynach, said there were many orioles. The height of an oriole nest varied according to the height of local trees. Average height of oriole nests in his study area was about 7–10m, with the lowest at 3m.

In southern Russia generally, orioles prefer high trees but in eastern areas of the Caucasus the habitat consists of dry steppe and desert. Trees there are those cultivated for fruit production and, as these are low trees, orioles are compelled to nest low. The birds eat a great deal of the fruit of mulberry, a species that has also gone wild in the Caucasus. Even so, local fruitgrowers do not regard orioles as a pest.

Siberia

The Siberian part of the Russian Federation stretches from the Urals east to the northern Pacific Ocean and Berents Sea. There are two species of *Oriolus* that breed, the Black-naped at the eastern extremity around Vladivostok and Sakhalin and the Eurasian Golden from the Urals to near Lake Baikal. In order to try to clarify things as far as this complex region is concerned, we shall move from west to east.

Figure 11. Breeding range of Eurasian Golden Oriole and *diffusus* Black-naped Oriole in Siberia.

Ural Province

There is very little descriptive information on habitat use at the western end of Siberia on the eastern edge of the Urals and around Ekaterinburg. The taiga forest habitat here and right through Siberia is as on the other side of the Urals: boggy, birch–conifer habitat, where the oriole is said to be common, but not so frequently found breeding (Ryabitsev *et al.* 2002). At the Urals University Biological Station in the grounds of the university itself, orioles are stated to be common in pine and mixed pine–birch woods, where a density of three to six birds per km square has been recorded. They are also present in copses and shelter belts between fields, but less common in boggy alder and bird cherry plantations.

Whereas the oriole breeds to around the 63° N parallel above St Petersburg in western Russia, here it is found no further north than 58° N, which coincides with the change of habitat from taiga to open, treeless country. In the years 1993–2002 in the Verkhnyaya Salda district, which is on the 58° N parallel, only four cases of breeding were proved, suggesting that this is the bird's northern limit (Ivanov & Pogrebnoi 2002). To the southeast of this region and near to the Kazakhstan border at Kurgan, the bird breeds in good numbers. Orioles have also been found breeding south of Omsk in a forest steppe habitat not far from Kazakhstan, but in the town itself it is described as a rare breeder.

Siberia Province

Further east in the Tomsk region but at approximately the same latitude, Blinova *et al.* (2001) claim that it is an 'everyday common bird', breeding in mixed birch, cedar and spruce forests; in copses in flooded or wet meadows; and in pine woods. It prefers light, deciduous woodlands along rivers, and grassy pine forests, city gardens and parks. It is very common in the Prichulymie Tomsk Chulym River area and in the suburbs of Makushino (Tarasov *et al.* 2002). Six to seven pairs were counted breeding in a 200ha game reserve rich in aspen–birch plantations and groves. Song was heard mostly from mid-May until the end of June. A song was recorded on 13 June 2005 at the southern edge of Tomsk, in the Summer Camp Garden (mixed woodland at 56° 24′ N, 84° 57′ E).

Further south, in the Novosibirsk area, there are two types of habitat: forest-steppe and steppe. Orioles breed in both, but numbers decrease the further south one goes.

In some parts of the Siberian steppe, there are very long belts of poplars, three to five trees wide, about a kilometre apart, described as 'reaching to the horizon', and probably planted as windbreaks. In June 2000, orioles were heard singing in them, though it is not known whether they were *O. oriolus* or *O. kundoo*, as both occur in that region (O. R. Marks & M. Marks, pers. comm.).

In the southern taiga, the bird was recorded in mixed semi-swamped and small-leaved forests. In subtaiga forests, it stays mainly in copses amidst fields. Fewer are

Habitat	*Individuals per km²*
Forest-steppe with birch–pine forests	22
Copses amidst pastures	14
Copses amidst fields, open forests, shrub-covered bogs	4–9
Small-leaved birch forest, peat, mesotrophic bogs	2–8
Southern Tobol river valley	14
Birch woods	14
Copses amidst fields	1–4
Birch–pine woods, pine plantations nr human settlements	<4

Table 16. Breeding densities in the Tobol river valley. From Blinova & Blinov, 1997.

found in small groves on low bogs and meadows, where they can sometimes be observed in aspen–birch forests. During the post-nesting period in July to the first half of August, the southern taiga is already devoid of birds, as the birds start to move into the middle taiga. In the subtaiga forests, numbers steadily decrease after the middle of June, and in the second half of August, all the birds have left the region. In the area between the limits of the southern subzones of west Siberia, the orioles seem to prefer open deciduous forests. Breeding density increases in the Ob River valley, especially in the floodland forests, reaching a peak in the area from the middle taiga to the subtaiga forests.Breeding density in various habitats in the Tobol river valley is shown in Table 16.

In the northern Altai the oriole is a common breeding bird, widespread throughout from meadow-steppe, low-mountain areas to light coniferous (taiga), middle-mountain areas. Post-breeding it sometimes ranges along cleared areas to the edges of dense coniferous, middle-mountain area, and sometimes also into subalpine meadows (Tsybulin 1999).

Habitat	*Pairs recorded*
Low-mountain, forest-steppe meadows with larch–birch copses	6
Ploughed-up fields in meadow-steppe, low-mountain areas	6
Mixed larch–birch forests	4–6
Small areas aspen–birch forests and tree-lined meadows	5
Middle-mountain, forest-steppe with open larch–birch forests	4
Birch–pine forests of the Katun river valley	4
Human settlements within these meadow-steppe landscapes	4
Coniferous–small-leaved deciduous trees in lower mountain areas	<4
Low-mountain, stepped meadows	2
Stepped meadows	2
Middle-mountain, park-type larch woods	1
Terraced valleys with fields, meadows, larch–birch copses	1

Table 17. Habitat preferences in the northern Altai. From Tsybulin, 1999.

In the post-breeding period, from the middle of July till the end of August, no changes in the distribution of the bird were observed, except that numbers were lower in all habitats.

Sometimes birds will appear in 'unsuitable' places, having bred elsewhere: for example, in clear-fell spaces in middle-mountain, larch–cedar forests and in larch–cedar forests with high-grass, subalpine meadows. In the middle-mountain part of the Siberian Altai region, where bird counts were repeated each year during five summer seasons, there were no notable year-on-year fluctuations of oriole numbers. Oriole breeding takes place in 78% of total areas of the middle-mountain areas, but in the remainder of the territory there have been no records, despite many years of searching. Orioles therefore are regular but not abundant, but it is thought that from 1979–83 in the middle-mountain part of the northern Altai oriole numbers remained constant.

In the northern Altai in average summers there were three pairs/km^2, in contrast to the north-eastern Altai, where there were fewer as a result of landscape differentiation. The bird is absent from the central Altai even though the habitat is identical to that of the northern Altai: extensive spaces in the middle-mountain, forest-steppe occupied by birch–larch forests. The total average summer count of orioles in the northern Altai is estimated at 24,000 individuals. The concentration is given in Table 18.

Teletskoe Lake is a remote lake in the Altai mountains surrounded by birch, pine and larch mixed forests and is fed by the Chlyshman and Bashkaus rivers. In the lower reaches of the rivers, orioles have flown through the high mountains to breed, and, according to Stakheev (2000), are common in pine–birch forests but rare in larch–birch forests around the lake.

The Golden Oriole's extreme eastern breeding range in Siberia is the Lake Baikal area (Ryabitsev *et al.* 2002). In the Irkutsk Region it inhabits the boggy forests of tributaries of the Angara river. The only confirmed breeding record so far is of a breeding pair in the lower reaches of the Irkut river in the vicinity of Vedenshchina village. Here on 2 June 1991, Ryabitsev found an occupied nest in the floodland aspen–birch forest with an understorey of bird cherry. He wrote:

> The construction, which took the shape of a hammock, was situated in the fork of a thin birch twig at 8 m above ground. The outside of the nest was

Habitat	*% use*
Forest-steppe, low-mountain areas	49
Forest-steppe, middle-mountain region	29
Meadow-steppe, low-mountain areas	10
Other	12

Table 18. Habitat use in the northern Altai.

covered by pieces of lichens. Male and female behaved a little warily and they stayed near the nest tree and attacked the Hooded Crows which were flying near. On 10 July in the nest there was one large feathered chick with down still on the head.

In Eastern Siberia two oriole species occur, i.e. both Eurasian Golden Oriole and the *diffusus* race of Black-naped Oriole. They are distinct in terms of distribution and numbers, and there is, as yet, little or no research into possible interrelations in the contact zone. The eastern border of the range of Golden Oriole goes through the Yenisei river valley, the western slopes of the Shan mountain ridge and covers extreme (marginal) western parts of Mongolia. In Predbaikalie (Front-Baikalia), it occurs throughout the Angara river basin and its right- and left-bank tributaries, excepting mountain regions of Prisayanie (Cis-Sayany Region). In the 19th century there were separate records of this oriole as far east as Usolie and Irkutsk.

Studies during the last half of the 20th century have found that the bird's range has expanded considerably near the eastern borders of its distribution. In Tuva, it was recorded in July 1958 in the floodland forest in Tes-Khem river valley (Flint 1984). Later it was determined that they sometimes breed in the northern part of Tuyinskaya hollow (Ubsu-Nur hollow). In western Sayany in 1987, it was recorded as a rare breeding species. Along the Yenisei river valley it occurs in riparian forests (one to two pairs/km^2), but away from there the species is very rare. On the border with Kanskaya (Kansk) forest-steppe, on the other hand, it is found only in the pine and birch forests. Along the lower Angara river and its tributary, the Taseeva river, in larch woods in old 'gari' (spaces left after forest fires) and cleared spaces, orioles are common, and at some sites even numerous (16–32 pairs/km^2).

There are records of Golden Oriole further east from the region of the Gadalei settlement in the Tulun district of the Irkutsk region. In the Chuna valley, surveyed in July 1988, population density between Nevanka village and Bunbui village was from 2.5 to 8 individuals per km^2. On high terraces between birch groves on the margins of open fields, numbers were considerably lower, at 0.7 to 1 per km^2. Higher along the river, as far as the Chuna settlement, only single birds were recorded, and the species was absent in pine forests above the banks of this river. The Golden Oriole is more common in the Chuna river floodlands, in sparse birch forests amid vast flooded meadows, and along channels with dense thickets of willow, hawthorn and bird cherry. It was noted also on islands in the middle of the Mura river (a tributary of the Chuna river) near to Chervyaka village, but over a total length of more than 100km it was recorded only twice. In both cases the birds were found in old birch forests on the edge of fields, near ravines with spruce trees. Along the Taseeva river, which begins after the confluence of the Biryusa and Chuna rivers, Golden Oriole is the most common species. The numbers are relatively high in this area because of the habitat; mixed arable and pastoral field shelter belts.

Much further east, Golden Oriole was found in July 1965 in the thin birch forest of the Elnik river floodland near the Buruk* settlement of the Kuitun district in the Irkutsk region. In conversation with Melnikov (1998), local people described it as 'a yellow bird the size of a starling and with an unpleasant call' (this must refer to the nasal-type contact call). It can sometimes be seen in flooded meadows of the Oka river, between Ust'-Kada and Okinsk villages (Kuitun district). 'This was confirmed by our own observations. On islands in the Oka River consisting of thin birch forests we recorded two orioles on 24–26 July, 1993. However, whether they breed here is not clear' (Melnikov 1998).

The birds often occur at Pleshi in the Biryusa river valley, where they undoubtedly breed, because their presence is recorded every summer. And finally there is reliable record of orioles breeding in the lower reaches of the Irkut river close to Lake Baikal, where a nest was found near Vedenshchina village in 1991. Probably, at the present time, oriole breeding numbers are low, but around Baikalia, according to Melnikov (1998), it is slowly expanding.

In Siberia we come to the eastern extremity of the breeding range of *O. oriolus*. Our correspondent in Irkutsk near to the southern shores of Lake Baikal, Igor Fefelov, explains that the population in the Irkutsk region varies from year to year as well as over longer periods, which equates with what appears to happen in the north-west of the breeding range. 58° N and 104° E are estimated to be the limits: to the east is the breeding range of the *diffusus* race of the Black-naped Oriole. There is no estimate of breeding pairs of Golden Oriole in this region, but counts of 2.5–8 birds per km^2 were recorded in 1988 between Bunbey and Nevanka along the valley of the Chuna river. In the note received from Melnikov containing this information, he says that as the birds were present all summer they must be nesting. The habitat is described as mainly birch and pine around clear-felled forestry, and in the marshy, denser forests where clearings occur. Nesting does not occur in the only pine species present, namely Scots pine, but does take place in both silver and white birch, which apparently often hybridise there.

Some riverine poplar is present, and there are small plantations of the hybrid cultivar *Populus balsamifera*. Along the rivers *P. laurifolia* and *P. suaveolens*, both wild species not found in Europe, are present, and Fefelov has found that the orioles use the latter species. Along the Goloustnaya river, the habitat differs in so far as the land slopes steeply and birch is interspersed with larch. Taken overall the birch is the most used nest tree. As with other places where the bird breeds at the edge of its range, numbers vary from year to year, and fluctuations also occur over the longer term.

The Golden Oriole's breeding range extends, just, into north-west Mongolia. This region is characterized by a sparse distribution of forests at higher elevations, wide mountain valleys in the Orkhon and Selenge river basins south of the Baikal

---

*Note: Buruk is a name of a Siberian nomadic tribe. They were translocated later to reservations east of Lake Baikal.

region of Siberia. Coniferous forests are found on the northern slopes of the region's mountains, while the southern slopes are covered with open steppe vegetation, which is a combination of forest and steppe flora, including species such as Scots pine and aspen. Both these species of tree are known to attract orioles here and elsewhere in Siberia.

TRANSBAIKALIA

The other *Oriolus* species to be found in eastern Siberia is the *diffusus* Black-naped Oriole. In south-eastern Transbaikalia (the region to the east of Lake Baikal), it was first recorded by local ornithologists in the 1870s, along the Aregun river in the vicinity of Stari Tsuruhaitui village. At that time it was quite a common species. Further studies have shown that *diffusus* Black-naped Oriole is more widespread in eastern Siberia than was first thought, but the breeding numbers are low everywhere. As a result, it was not discovered by many of the researchers working in Transbaikalia. At the present time two new finds are considered trustworthy. The first is on the Ulan-Burgasy mountain ridge (eastern Transbaikalia) on 4 June 1975 (Dorzhiev 1996). The second is a record from the Solzan river valley in the Baikalsk area on 10 June 1985, where a male was collected (Melnikov 1998). The last find shifts the border of the range of this species far to the north-west. It is possible that within the limits of the Irkutsk-Cheremkhovskaya Plain there is a contact zone between the orioles, though both species are extremely rare here. In the opinion of the report authors, it seems that at the present time there is a gap between the ranges of these species, but this does not exceed 80km.

Considerable thinning out of forests in the south of eastern Siberia, caused by virtual clear-felling and logging of the pine forests and by fires, creates preconditions for expansion of ranges by both oriole species. The regions of the last records of Golden Oriole (Chunsk, Nizhny-Udinsk, Tulunsk, Kuitun and Irkutsk districts) are the most likely. In cleared spaces, the river floodlands are the most typical microhabitats of orioles in the Irkutsk region, and these are surveyed regularly, so the possibility that either oriole species will be missed is extremely low. This enables modern data to be used with confidence as a base for tracking the changes in distribution and number of both oriole species in the south of eastern Siberia.

The Golden Oriole is a vagrant on the south-west shore of the Krilyon peninsula of Sakhalin Island, having been recorded three times. In the Shebuninka river valley on 13 June 1971, feathers of a male were found, and on 27 June a probable female was also found in flooded alder–willow, open-forest habitat. In Kuznetsovka river valley on 27 June 1974, a female with worn feathers was caught. Sakhalin is north of the normal breeding range of the Black-naped Oriole.

CHAPTER FIVE

The special case of Kazakhstan

Kazakhstan is a very large country of 2,717,300km^2, stretching from the Caspian Sea to western China – an area as big as Canada or the whole of Western Europe, with a human population of 17 million people, living predominantly in the towns. This is a land of great open and uninhabited spaces. Sandy deserts and extensive grassy steppes occupy the largest part of the country, with large inland lakes such as Lake Balkhash and the salty Aral Sea. On the south-eastern border are the Tien Shan Mountains, which are the spine of central Asia, extending from the northern Himalayas, and are the source of water that feeds lakes to the north. Endemic trees include a birch, *Betula talassica*, and a poplar, *Populus talassicus*. In the Turanga forest another endemic species of poplar tree, *Populus diversifolia*, occurs along the banks of Kazakhstan's rivers and inland deltas. To the north, such forests are found along the Ili river, which flows through an otherwise desert landscape. These forests provide good habitat for many species of birds, and especially for orioles. The capital, Almaty, is an abundantly leafy modern city with a backdrop of the Tien Shan Mountains. Aksu Dzabagly Nature Reserve, the oldest reserve in central Asia, established in 1927, is not far from Almaty and conserves some of the botanical habitats of the Tien Shan Mountains 1400 species, many of which are rare or endemic to the western Tien Shan.

Kazakhstan is unique in that it is the only country where both *O. oriolus* and *O. kundoo* breed in large numbers. The Indian Golden Oriole, *O. kundoo*, only breeds in the south and east, mainly around Almaty and into the lower reaches of the Tien Shan mountain range to about 1880m. It also breeds along the Syrdaria river valley as far as Kazalynsk. Those that reach the northern shores of the Aral Sea, into which the Syrdaria river flows, are on migration.

Eurasian Golden Orioles in Kazakhstan breed mainly in the west, north and east of the country in the taiga, which are forest galleries in the floodplains of central Asian and Siberian rivers. They often consist of impenetrable thickets around damp, grassy clearings. The taiga represents a virtual oasis for both resident and migratory wildlife. Main flyways for migrating birds pass through the water bodies in the riparian forests and these serve as wintering sites, stops along the migration routes, and sites for colonial nesting. These riparian forests have been severely threatened by man's activity for many years now. Restoration to prime condition is a priority, but at least orioles have found a home here. Tugai, a type of habitat consisting of the tugai communities described below, lines the shores of central Asia's largest rivers, the Syr Darya, Amur Darya, Murghab, Tedzhen, Ili, Zeravshan, Vakhsh, and the Naryn. The tugai plant communities comprise poplars (*Populus diversifolia, P. pruinosa*), dzhidda, willows and tamarisk forests, which

Figure 12. Breeding range of Eurasian Golden Oriole *O. oriolus* and Indian Golden Oriole *O. kundoo* in Kazakhstan.

alternate with meadows and reeds. The biodiversity of the tugai ecosystem is very rich, and includes a wealth of invertebrates, making it an excellent area for orioles. Here the poplars are not the familiar European species but endemic to this and neighbouring states like Uzbekistan and Kyrgyzstan. It should be stressed that for many regions of Kazakhstan reports of breeding have not been confirmed, and could refer to non-breeding birds summering in the region.

In the south of the country, both species pass through the Chokpak ringing station on both vernal and autumnal migration but at differing rates. In spring during the period 1969–2005, 101 *O. kundoo* were ringed as against only 51 *O. oriolus*, whereas in the autumn, the situation is reversed: 28 *O. kundoo* ringed as against a remarkable 548 *O. oriolus*. Edward Gavrilov (*in litt.*) explained that this difference occurs because in spring the desert areas are green, with many flowers and insects, which the birds take whilst moving through to Siberia and northern Kazakhstan and only divert east due to any adverse weather conditions. In autumn, on the other hand, they migrate through forested mountains (which is where Chokpak is situated), thereby avoiding parched deserts. In other words, it is a local loop movement caused by differences in ecological situations in spring and autumn.

The Eurasian Golden Oriole breeds from the west of Kazakhstan near the Russian border, but not by the Caspian Sea, and in a wide strip along the border with Siberia to the north and east. Within this area they breed exclusively in deciduous forests such as birch groves and oak groves of the forest-steppe, aspen forests, poplar–black poplar thickets in river valleys, and willow and karagach groves. It inhabits old field shelter belts and is also found along main roads and in gardens. In parts of northern Kazakhstan, where it breeds exclusively in pine woods, Eurasian Golden Oriole does so at a rate of 14 and nine individuals per km^2 in the northern and southern sub-zones respectively. In old groves of semi-abandoned

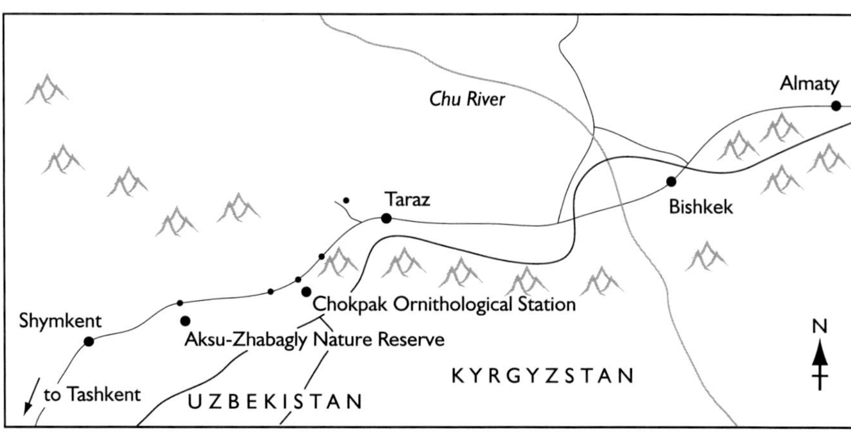

Figure 13. Location of the internationally important ringing station at Chokpak, Kazakhstan. After Gavrilov, 2007.

settlements the rate is four per km^2. Nest-building by a pair of birds was observed on the 28 May, i.e. the Eurasian Golden Oriole begins breeding just after arrival (Blinova & Blinov 1997).

The Kazakh forest-steppe is quite different from the similarly named habitat in European Russia. It is some 300–500km further north of the equivalent landscape in European Russia, has a more continental climate, and is flatter with a much higher proportion of wetlands. The northern border of this ecosystem stretches from the southern Ural Mountains along the southern end of the West Siberian Plain, and south of Novosibirsk (in Siberia) it continues to foothills of the Altai and Sayan Mountains. The southern edge goes along the Russia–Kazakhstan border, mostly within Russia, but near Petropavlovsk it comes further in to eastern Kazakhstan and reaches the Middle Irtysh river. The width of the forested steppe zone is 150–250km and forms a ribbon along the border of the two states (Gvozdetskiy & Mikhailov 1987).

In central Kazakhstan, the Eurasian Golden Oriole breeds only sporadically, though in summer there are occasional sight records. This steppe region is the country's vast plains area, where there are very harsh sandy conditions. There are some areas of poor grass such as near Shortandy (close to Astana), where agricultural research is carried out. Here, our species is a rare breeder in the nearby poplar forest belt. This kind of steppe was typically grazed annually before about 1900, although the study site itself is not grazed. It represents one of the last few square kilometres of virgin dry steppe in northern Kazakhstan, in an area largely converted to wheat production in the 1950s. The remnant virgin steppe is thought to be in good condition, with large amounts of carbon stored in its soil. A singing male was once sighted on 20 July in birch groves amidst the hilly steppe between Shortandy and Baryshevka, but no breeding was proved.

In some tall pine woods near Amankaragai on 18 June 1957, 20–30 birds were seen to concentrate on a group of trees badly infested with Birch Beetle-sawyer. During passage the birds form small flocks, consisting in rare cases of up to 30 birds. Generally at this time, however, they are mostly single or in small family groups.

In the central Kazakhstan steppes the birds search for caterpillars from the air, stopping, hovering over one place, and swiftly dropping to the ground. Having caught the prey item, they sit with it on the ground or on a small stem of a grassy plant (Prokop, 1983). The main food here is insects of large and middle sizes, predominantly lepidopteran caterpillars, including hairy caterpillars. They also take beetles, bugs, grasshoppers and sometimes molluscs. In autumn, they feed on different fruits, such as cherries, mulberry, cultivated plum, and also the wild plum *Alycha*, which grows in gorges and on northern slopes of mountains up to 2,200m. They have also been recorded feeding on rotting apples and pears.

After breeding and during passage, our orioles occur in more diverse sites, such as the younger forested areas, thickets of the pea tree (known locally as the karagana), tamarisk, various *Nitraria* shrubs, spreading juniper thickets, in isolated trees, in vineyards and sometimes in dense stands of reed. The use of the pea tree, which is in fact a tall shrub, is interesting, in that the fruit is just as a normal pea in a pod. The orioles have obviously discovered how to open the pods to extract the fruit.

Indian Golden Orioles, on the other hand, are only to be found in the south and south-east of Kazakhstan, where they inhabit mainly plains and foothills and mountain river valleys. In the Tien Shan they can be found up to 1700m, more rarely to 2200m. In Altai, they are not found above 1100m, but when on migration they fly considerably higher, for example in Altai up to 1800m and in the Talass Alatau region even as high as 2700m. Only *O. kundoo* occur in and around the capital, Almaty, where they frequent gardens and poplar plantations along river valleys. Their distribution in Kazakhstan is summarised by Edward Gavrilov (*in litt.*):

> *Kundoo* breeds in the Tien Shan foothills and in the Syrdaria valley up to Kazalinsk, but on the north shore of Aral Sea and on Balkhash Lake it is a passage migrant only. In the central plains of Kazakhstan small groves of trees exist but I think birds do not summer there, and late records belong to lingering migrants only. They can also be found near Jongolsay in the Ural valley, north of Atyrau.

It appears that *O. kundoo*, after wintering in India, arrive in spring about a fortnight earlier than do *O. oriolus*, which are arriving from Africa. First appearance dates are recorded at the end of April, and by the end of May nests with eggs can be found (as early as 28 May) (Gavrilov 1974). In northern Kazakhstan Eurasian Golden Orioles begin to breed approximately two weeks later, at the end of May to the beginning of June. They normally lay one clutch per year, but in the case of the loss of a nest with eggs, repeat breeding does occur.

In Almaty Indian Golden Oriole is one of the commonest breeding birds of the town, although it is not evenly distributed, being at a lower density in the south and south-east of the town. In the northern suburbs, in 1961, breeding pairs were recorded in two blocks of human habitation, and in the more open conditions of Gorky Park Indian Golden Oriole nest at about 200–300m apart. In the central and especially in the northern part of the town breeding is more patchy. Such unevenness of the nest distribution is explained by different species of tree plantations. The extent of humidity is important too. Water is more abundant in the upper part of town, so vegetation here is more luxuriant and lasts longer. In the northern part of the town it breeds only at a few sites, for example in cemeteries and in the park at Almaty I Station. In Almaty, a nest was found on 28 May with two eggs, another with three on 29 May and one with four on 1 June. Three nests with almost fresh eggs were discovered on 28 May back in 1916, thus giving an indication of reasonably early starting times. In a footnote to his paper on the birds of Almaty, the author Borodikhin (1968) remarks that 'the Golden Oriole is very beautiful. This bird sings melodiously and soundly; its presence decorates town.'

There is an interesting remark in the *Birds of Kazakhstan* (Gavrilov 1974) to the effect that because of the nature of their feeding, orioles are one of the most useful birds for forestry, because they exterminate a variety of pests in forest plantations, including hairy caterpillars, which are not eaten by all birds. In this context, particular mention is made of orioles feeding on larvae of the Brown-tail moth, which are very hairy and can rapidly defoliate plants.

In conclusion, it is clear that Kazakhstan is a very important country for both species of golden oriole. Fortunately, its ornithologists have undertaken a great deal of research, which has contributed greatly to our knowledge.

CHAPTER SIX

Climate and weather

CLIMATE

It is a truism to say that climatic changes have an enormous influence on the distribution and population dynamics of flora and fauna. The colonisation of southern England by animal species typical today of southern Europe probably began in Saxon times, and reached its peak in the medieval warm period known as the Little Climatic Optimum. Burton (1995) notes that the Anglo-Saxon scholar Aelfric was familiar with such continental birds as the Hoopoe, Golden Oriole, Eagle Owl and Quail. Furthermore, W. H. Hudson (quoted in Nicholson 1950) argues that the fact that the Golden Oriole had a local or dialect name, 'woodweele' or 'wodewele', supports the view that it was a familiar bird in the Middle Ages. He quotes a contemporary poem by an unknown author:

The woodweele sang and wolde not cease
Sitting upon the spray
So lowde he wakened Robin Hood
In the greenwood where he lay.

Similarly, Chaucer, in the 'Romaunt of the Rose', wrote of the 'woodweele' or 'wodewele' in a way that suggests it was a well-known species: 'in many places were Nyghtyngales, Alpes (Bullfinches), Finches and Wodeweles, that in hir sweet song delyten'. Hudson's assertion that the woodweele is the Golden Oriole has been challenged by some authorities, but Nicholson is convinced, observing that the name 'woodweele is suggestive of the unmistakable song of the Golden Oriole, which is variously described as 'weela-weeo' or 'aweela-wilee'. Moreover, the onomatopoeic Dutch name 'Wielewaal' is strongly suggestive of a common origin with the old English name.' However, the fact that Chaucer may have heard or even seen orioles does not necessarily mean that they were breeding. We now know that the species is a great wanderer, especially males in their first two years.

The centuries of the Little Climatic Optimum gradually gave way to a period known as the Little Ice Age so that, by the end of the 17th century, when this period was at its peak, those typical Mediterranean species like Hoopoe and Bee-eater no longer, or only rarely, bred in Britain. This mini ice age gradually receded, so that by the mid-19th century our region was again in a warm climatic phase, which, while not as hot as the medieval warm period, encouraged a northward expansion of the Golden Oriole's breeding range in north-west Europe. Denmark, for example, was colonised from northern Germany from the 1850s; by 1976 the breeding population there was estimated at 200–400 pairs. By the later decades of the 20th century, therefore, the bird had also established itself, or more correctly re-established itself, in Britain, notably in south-east England and in the fenland basin of East Anglia, with sporadic breeding records as far north as Scotland. Essentially, the Golden Oriole thrives best in drier climates with warm summers. Thus it is able to breed as far north as south-east Finland with its short, hot summers, but tends to avoid the cooler, maritime summer climates of north-west France (Brittany and Normandy), Norway and the British Isles, except when the climate alters towards a drier continental type.

Any discussion of the effects of global warming and concomitant climate change is so fraught with caveats and hypotheses that it is impossible to make any categorical statements about the effect, if any, on our study species. What is clear is that the process of desertification is accelerating to the north and south of the tropical zones, with consequent effects on flora and fauna. For example, the Sahel region to the south of the Sahara, which constitutes the winter feeding grounds for migrant insectivorous species such as flycatchers, wagtails, warblers and shrikes, is drying out. The result of this desertification is not only that the birds' winter food supply is reduced, but that their trans-Saharan migration is made longer and more difficult. Orioles are likely to fare rather better in their winter quarters and on migration, in that they winter further south in Africa than the Sahel. British orioles winter in arboreal habitats in Zambia, Zimbabwe, Malawi and the Congo, having followed certain defined routes as opposed to indiscriminate flights over desert areas. On these routes, the orioles have been able to stop off to feed, whereas trans-Saharan autumn migrants do not have this opportunity as many oases will have dried out by this time. Orioles feed mainly on the abundance of fruit which is available at this time, especially figs, which are known to mature at different times throughout their range.

It is only on its spring migration, largely over the Sahara, that the oriole suffers to any degree. Although it stops off in numerous wadis and oases, it does have problems gaining sustenance, as evidenced by the number of dessicated carcasses that are picked up throughout the sandy wastes. Once they attain North African coastlines, where there are gardens to feed up in, the orioles are safer and ready to cross the Mediterranean, which is less of a hazard to them. The southward autumn route is more strictly defined and mainly follows routes that enable them to find fresh fruit nearly all the way.

Less certain is the effect of climate change on their food supply on their return to their breeding areas in spring. Current research indicates that the insect life on which they feed is emerging earlier from its winter state as egg, larva or adult. British lepidopterists note that overwintering larvae are vulnerable to attack from viruses and fungal moulds, which thrive in the warm, wet conditions of recent winters, with an inevitable reduction in food supply. Conversely, eggs laid in autumn by other insect species seem to be emerging earlier than previously, producing small larvae which provide a good food source for tits, Robins and Wrens, and for short-distance migrants like Chiffchaff, Blackcap and pipits, some of which have started breeding earlier as a consequence (Sanderson *et al.* 2006). Because orioles also feed on larger and less palatable prey than the smaller passerines, they should have a slight advantage.

WEATHER

While it is difficult if not impossible with our current stage of knowledge to make predictions about long-term trends resulting from climate change, we can be more confident in identifying the effects of fluctuation in weather. To take one dramatic example; in 2001, temperatures in the fenland study area were well below average and were further perceptibly reduced by strong, persistent, force 3 north to northeast winds from early March (for 23 out of 31 days) to April (for all but two days), carrying on into May for about seven days, causing a wind-chill factor enough to severely retard leafbud even on early-leafing varieties of poplar, so much so that whatever insects there were to lay eggs on emerging foliage were unable to do so. In turn the orioles that arrived back from migration found trees with little or no foliage on which to forage, thereby making it impossible for them to achieve the necessary protein levels to mate and to produce eggs. As a consequence, half-built nests were abandoned; the birds left their territories and began wandering over large areas searching for food or, in some cases, moving away from the area. There was one exception, where the only successful breeding occurred. In this case the nest was placed in deep woodland towards the southern leeward edge, where a degree of protection from the inclement conditions presumably afforded a favourable microclimate. Given that a similar cold spell also occurred in Denmark (G. Fahlberg, pers. comm.), in north-west Germany and The Netherlands, at least, it is reasonable to assume that there was a knock-on effect on the East Anglian population in subsequent years, given that any recruitment from the continent would be adversely affected.

Orioles seem to be particularly vulnerable to day-to-day changes in the weather, especially in May when the birds arrive, in June during incubation, and during the period when nestlings are being fed. In 1987, the first full year of the Golden Oriole Groups' co-ordinated survey work, Jez Dagley (1987) found that many nests contained dead nestlings in distressing numbers. In late June of the same year, several days of persistent rain and cold again inhibited the parents from food gathering or, if they did, then the young were exposed to the wet with the consequent risk of hypothermia and pneumonia. At the time when the nests were inspected, 24–26 June, the chicks were six to eight days old. Between 18 and 25 June, the total rainfall was at least 27.1mm. This was exceeded only in 1997, when the figure was 34mm, and 20.8mm fell on a single day (27 June), when the birds were close to fledging or had fledged and were probably safe. Apart from these extreme conditions, the average annual rainfall for the same period annually in the fenland is 6.3cm. In 1995, it was not the total rainfall that was the problem; on one day, a concentrated group of almost tropical storms passed through the Fens. As they travelled, high winds, heavy rain and intense lightning, often pink in colour, destroyed many nests, and the adults were later to be seen wandering around aimlessly. One female with a brood patch was eventually trapped and ringed near Felixstowe, on the east Suffolk coast, on 23 June, at a time when she would normally be rearing a family. It is ironic that in 2001 there was no rainfall in the critical period, but by that time there was only one nest with young. In the record hot summer of 2003 breeding numbers picked up after the very poor seasons in 2001 and 2002. Birds arrived roughly on time to find foliage in a good condition and plenty of invertebrate food available. Five definite and one probable pairs bred, producing a good crop of young.

In most years, if the overall temperatures are suitable and the periods of rain short or just showery, adults are able to forage adequately enough to rear reasonable numbers of young. An intriguing question is whether the long, hot, dry summer of 1976 increased numbers of orioles in fenland in subsequent years. RBBP reported higher numbers: in fact, the highest estimated number of pairs nationally occurred in that year, followed in 1977 by the second highest.

Summing up, we can say it is too soon to talk about the effects of long-term climatic trends, but that, in Britain at least, the birds are at the mercy of the vagaries

Fledgling production per nest	*June weather*				
	Warm	*Cool*	*Dry*	*Wet*	*Total*
>1.39	21	11	21	11	32
<1.39	9	26	12	23	35
Total	30	37	33	34	67

Table 19. Fledgling production in different climactic régimes. From Millwright, 1997.

of our weather, and therefore, not surprisingly, maintain no more than a toehold in this country.

Milwright (1997) analysed the annual fledging rates of fenland orioles plotted against weather data (rainfall and mean temperature), and confirmed that productivity was lower in years with a cool, wet June, and higher in years when June was warm and dry. As, however, the month of June has every combination of temperature (cool/warm) with rainfall (wet/dry) during the study period, it is best to study the two parameters separately. The sample sizes are too small to produce statistically significant results, but Milwright came up with an ingenious method of producing meaningful statements, by plotting high and low fledgling production against the four weather variables (Table 19).

The positive relationship with June mean temperature is highly significant ($p<0.01$), as is the negative relationship with June rainfall ($p<0.05$).

CHAPTER SEVEN

Courtship and nest-building

The Golden Oriole Group data recording the first-date arrivals of males and females on site (based on first recorded dates) are summarised in Table 20.

At the core Lakenheath site, PD recorded arrival dates for the years 1990–2006, which gave a mean date of 5 May (±2.7 days). In 2005, NS noted the first bird at Lakenheath on 30 April, suggesting that arrival dates have become earlier in recent years. These dates are about a fortnight later than mean arrival dates calculated for France and The Netherlands. After pairing and site selection, we have a mean start

Sex	*Mean arrival date*	*Range*	*No. of observations*
Males	14 May	1 May–2 June	47
Females	21 May	8 May–3 June	18

Table 20. Mean arrival dates at the fenland study site.

for nest-building, based on 11 observations in the study area, of 26 May (range 25 May–8 June). Duration of nest-building varies with conditions, from five days in good weather to nine days or more in bad.

PAIR FORMATION AND COURTSHIP

In the East Anglian fenland basin, where most of the British birds breed, the first males generally appear on site in early May, although earlier arrivals are being recorded in recent years, and on a few occasions, females arrive soon afterwards. Those early females are usually senior birds and the ones that have used a particular site before. One particular male is known to have used a site for up to 10 years. In that time he had at least two and probably three partners. One partner was herself very senior in appearance but was lost to him two years before he himself no longer returned, presumably dying at a ripe old age. Whether senior females, having lost a partner, continue to return first to a site before mating with a young wandering male previously siteless is probable but not as yet confirmed. It is thought that the female who lost her 11-year-old mate returned to her site, but because of females' more reticent behaviour on lone arrival it is not known if she came before finding her new mate.

Once a male and female form a pair, there is much chasing around and through trees and across open country, with a lot of aerobatic twisting and turning. The male closely pursues the female, even appearing to nip her tail feathers with his beak. Sometimes a second male, presumably an immature non-breeder, joins in. When this happens, singing by both males is strong and intense, with many variations on the 'weela woo' phrase most commonly used. Also at this time falcon-like calls can also be heard, giving rise to an excitement that can even affect a human observer with its intensity. A description of courtship behaviour at a site in eastern Germany records how a pair moved excitedly all around the branches of a tree, with both male and female calling and singing, the female's song somewhat subdued. Song was interspersed with squawking calls. Tail feathers were spread and wings fluttered, as the male especially adopted all manner of postures to impress his mate (Höppner 1954).

If, having set up a territory, a male fails to attract a mate, he will wander off after two weeks or so, no doubt to become a threat to other territorial males. Again intense, variably phrased singing will take place, with attendant flight action, aimed at removing the intruder. The female joins in, making cat-like rasping calls, which males also make at such times. It can be seen that variation in song and calls constitute an important part in territorial defence. Courtship proceeds over a period of up to seven days. A report from Bavaria, Germany, supports this observation (Reinsch and Warncke 1971), although they also thought that the pairing up of late arrivals generally happens over a shorter period. Studies in Picardy, France (Etienne and Montel 1986) estimate that courtship and nest-building together can take up to three weeks, which is also true for sites in the fenland study area. In Switzerland, a female was observed to start nest-building on the second day after her arrival, whilst courtship continued (Buxton 1928).

Many other species of birds engage in courtship feeding, but we have no record of this in our study area, nor any reference to it in the literature. While one cannot be sure of everything that occurs in dense-canopy foliage, often at great height, other observers are of the opinion that Golden Orioles probably do not practice it.

Copulation mostly takes place in well-concealed places. Feige (1986) has observed a male uttering his 'weela woo' song whilst hopping around a female who has her tail fanned to one side in preparation for mating. In this instance, the male mounted her up to four times in succession. This was observed in a tree, but there is a record of orioles apparently mating on the ground, which must be an unusual event (Gross 1899).

DEFENCE OF TERRITORY

Territory sizes are hard to define and vary according to the layout of the habitat. Pairs are reasonably tolerant of each other and overlap can occur during foraging forays. An exercise carried out in a shelter-belt situation where trees ranged from two rows to five rows showed that approximately 400 poplar trees constituted the territory. The smallest distance between two nests to have been recorded in the study area was 300m. See Etienne and Montel (1986) for territory sizes in Marquenterre (Somme) region. Baumann's telemetric research in Schleswig-Holstein 1991–95 (Baumann 1999) casts interesting light on the nature of territoriality in orioles and, in particular, the difference between defended territory and song territory. A total of 12 orioles was trapped, ringed and fitted with transmitters. The data from eight males from different breeding pairs were used (see Table 21). The size of the general area is very variable, the extent of each being

Pair	*Bird's range (ha)*	*Song territory (ha)*	*Defended territory (ha)*	*Number of soundings*
Pair 1	60	17.2	1.2	121
Pair 2	76.6	2.1	1.8	82
Pair 3	94.6	6.8	1.1	230
Pair 4	8.3	14.7	1.1	100
Pair 5	6.8	17.3	0.7	115
Pair 6	23.3	20.8	1.1	125
Pair 7	36.7	61.9	1.8	118
Pair 8	20.6	44.4	2.1	276
Mean	40.9	23.2	1.3	
SD	14.4	7.1	0.2	

Table 21. Range analysis of eight radio-tagged male Eurasian Golden Orioles. From Baumann 2000a.

determined by topographical features. This in turn determines the bird's range within each area, again very variable. Perhaps most interesting is the difference between the fairly restricted area which the bird defends territorially, i.e. the immediate vicinity of the nest, and the much greater area over which the bird roves to announce its presence in song. The considerable variability in the extent of song territory is also accounted for by topography and in particular by vegetation types. In some cases (Pairs 4, 5, 7 and 8), it might appear strange that the size of the song territory is greater than the size of the bird's range within the area, but this is accounted for by the nature of the definitions of the two areas, in that all occurrences in the song territory, however distant from the nest, were counted, whereas in the case of the bird's range, a quarter of the most distant records were discounted.

Intraspecific rivalry

Studies in Azerbaijan noted that birds arrive in pairs and just after that begin to select the nesting area but that the boundaries of the nest site were not rigorously defended. Quite often a singing male would stray into a neighbouring territory but would not stay long. There were some aggressive contacts at this time, but they were sporadic, even though birds often nested not more than 130–150m apart. It was only within the period of egg-laying and incubation that aggression was noted. The parent birds rarely left the nest site, and if other orioles did come into their territory, the male flew very aggressively and quickly chased off the intruder.

While studying nest-building near Geneva, Buxton (1928) noted an interesting interlude when a neighbouring pair visited the nest he had been watching and described it as follows:

> All work ceased, the four of them sat in a little oak tree, hens on the top, cocks on lower branches. The cocks were extremely voluble and at intervals quarrelsome, but the hens sat in dead silence with their bill tilted up, looking at the view. Altogether it was an absurdly human performance, at which all four birds seemed bored to death when the visitors suddenly took themselves off and the home pair resumed work building.

This certainly shows that the birds are not so determinedly protective of their territories. Similar neighbourliness has been recorded in France (Etienne and Montel 1986).

Observations at our core site at Lakenheath revealed that same-species intruders probably arrive in the early daylight hours and may persist for several days in the case of a determined individual. Non-breeders can stay in the area all season, which leads inevitably to repeated antagonistic forays. PM watched the determined chasing of intruders well away from the actual nest site over open arable land, often up to 0.5km distant. This was before the resident pair had finished building their

Behaviour	*Reference*
Singing males in slow flight with jerky wingbeats	Reinsch 1959; Feige 1986
Birds falling to the ground with claws interlocked	Gromier 1909
Threat postures (spread wings, ruffled plumage, etc	Reinsch & Warncke 1971
Physical fighting	Feige 1986
Physical fighting with backdown and retreat	Etienne & Montel 1986

Table 22. Nest-defence tactics.

nest. Calling ferociously, the territorial male will establish himself on a high-lookout tree, ruffling his feathers menacingly, often accompanied by bill-snapping and excited hovering whilst still clutching a twig with one foot. Other behaviours have been noted by various authorities (Table 22).

One interesting record is of four males, comprising two immatures and two adults, facing each other and whistling, alternately facing their opponents, for approximately one hour's duration (Boddeke 1986). This occurred in July amongst the Spanish Pyrennean mountains. The date would indicate that this was post-breeding or at least family-party antagonism. In contrast to the foregoing, Dagley (1987) mapped several territories at Lakenheath, and showed that there was overlap, at least when birds were foraging.

NEST SITE SELECTION

Research carried out by Natalie Fleming in 2005 (Fleming 2006) sheds useful light on the site preferences of orioles, particularly in respect of the selection of edge habitat. Her conclusions are worth quoting in full:

> Throughout the results, perimeter appeared significant . . . Milwright (1997) found 80% of nests to be located within 2 trees of the edge of a site, therefore suggesting that edge habitat is specifically chosen. It is known that shelterbelts are often used without any effect on breeding success (Dagley, 1987; Dagley, 1994), There has been some discussion as to why a large poplar plantation such as Lakenheath supports fewer territories than theoretically possible (N. Sills, pers. comm.). If the importance of edge habitat is indeed highly significant to Golden Oriole site choice, then this could be a limiting factor at large plantation sites. Golden Orioles also display a preference for early leafing poplar varieties. Moreover, research carried out by Mwangi (2006) at Lakenheath revealed that trees at the plantation edge may come into leaf earlier than those in the centre. If a male Golden Oriole is to maximise his breeding success, particularly in the midst of competition, territory site choice would be important. Sites with a lot of attractive edge habitat would give the best choice

of nest sites to potential mates. Plantation edges and shelterbelts may also contain a higher proportion of Lepidoptera which may benefit from the structural tiers of vegetation that edge habitat provides. Surrounding agricultural fields may provide an increased nectar resource for Golden Oriole prey species and hence increase a specific food resource for chicks during the breeding season. These factors provide evidence that edges and shelterbelts are potentially better for Golden Oriole breeding (as suggested by Dagley 1994).

The selection of the precise place to site the nest is often determined by the birds' previous choice – at a regularly used location. The nest is generally built in a tree which is close to where the pair had nested in previous years. A strong indication of a new pair, or part-pairing, is when the nest is placed in a new part of the territory, as instanced by the Cambridgeshire site, where a succession of changes of partners resulted in the selection of different areas within the plantation. The site, which was approximately rectangular in shape, consisted of a 2ha plantation and two separate, double, poplar field shelter belts, with a smaller triangular 30-tree plantation at the farthest point from the larger example. Distances involved were 1–2km. In 1987, when first found, Male A, a ringed bird now known to have been at that time a second year, was accompanied by Female 1, quite a senior bird judging by her plumage. During that pairing, all nests were built in the larger plantation. When Female 1 disappeared, Male A found a new, younger mate, Female 2, and for the next five seasons, the nests moved to the smaller plantation at the far end of the site. Female 2 vanished for whatever reason, after which the nest placings returned to the original wood for two years, with Female 3. When Male A no longer returned, having survived 11 years, Male B took over the territory with Female 3. Nests were in the vicinity of the original Male A + Female 1 pairing, but in the last year of breeding recorded at this territory, the nest was in a completely different section of the larger plantation and was assumed to be that of yet another female, Female 4, because the pattern of behaviour of feeding young was very different.

In summary the network of pairings was as follows. No confusion of observation is possible because there was never more than one nest in the territory in any single year.

Male A	>	Female 1
Male A	>	Female 2
Male A	>	Female 3
Male B	>	Female 3
Male B	>	Female 4

When deciding on the exact place to build her nest, a female will inspect various forks and test for strength and ease of approach and departure (Reinsch and Warncke 1971). Other observers, e.g. Feige (1986) and Wassmann (1990), believe

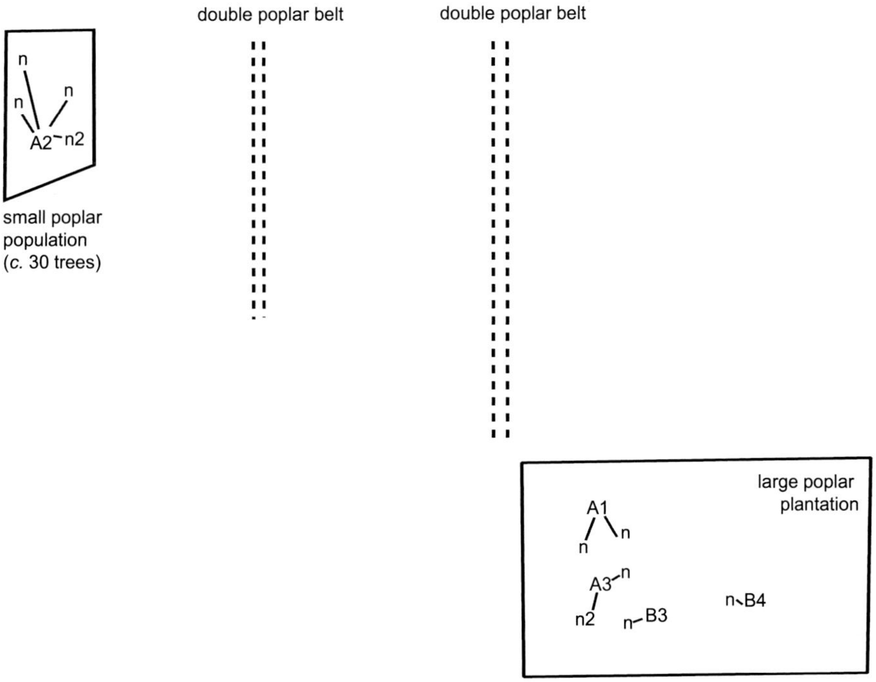

Figure 14. Nest placements at a Fenland site over 12 years. The data suggest that females choose the actual nest site while the male selects the overall site. Male A nested with three different females (1, 2 and 3); male B nested with two females (3 and 4). Female 3 was the only one that mated with both males in turn. There were never two nests in any one year. n = nest placement, n2 = nest used twice (in separate but not necessarily consecutive years).

that females will change to another fork if they have difficulty in fastening nest materials, or if they appear to be unhappy with the approach and departure routes.

After a tree has been selected, the nest is almost always placed approximately one-third of the way down from the crown of the canopy. The Golden Oriole Group has found that the height of the nest is always influenced by the height of the tree.

There was a notable exception that took place before the Golden Oriole Group embarked on its own survey. Ron Hoblyn (pers. comm.) described how he found a nest in an apple orchard (in what was later part of the study area). As the nest was no more than 1.3m above the ground, he was able to look into it. In 1995 in Eastern France, there is a record of a similar instance in an orchard, with the nest placed at exactly the same height (P. Etienne, *in litt.*). On that occasion it was known that the pair were likely to be first-time breeders, because the male was still green. Etienne's experience when researching nest heights in the Somme region of France was different in so far as most nests were in the bottom third to half of the tree's elevation. Actual heights recorded by him are in Table 23.

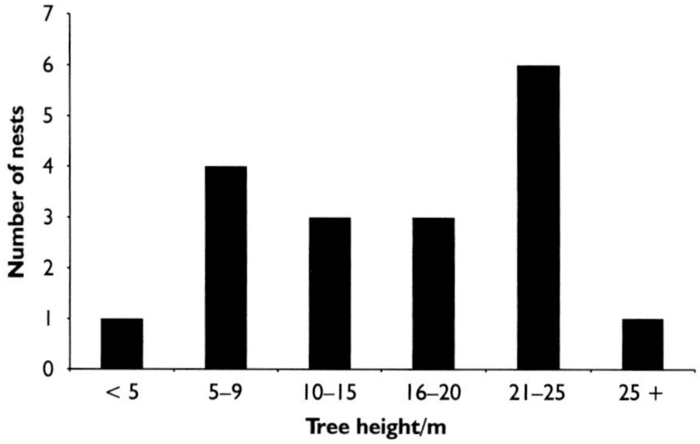

Figure 15. Range of oriole nest heights in poplars in the Fenland Basin study area.

Tree species	*Nest height (m)*
Poplar	14
Alder	6
Birch	8

Table 23. Nest heights in the Somme region. P. Etienne, pers. comm.

Having said that one has to remember that, in the main, birch trees are not as high as most others. Some data concerning the height of nests are given Table 24. There are further records to be gleaned from Ganya (1969), Georlette (1939), Labitte (1951), Melde & Melde (1977), and Zippelius (1972) on this matter, but all seem to point to a similar outcome.

Further investigation has shown that, on any given branch, which is normally near horizontal or at a slight upward angle, the nest is placed in a fork in thin branches approximately 0.5–1cm in diameter. One branch is the main stem of the bough and the other a sub-branch. Although Feige (1986) and Zippelius (1972) have seen nests close to the trunk, most are roughly two-thirds of the way out from it, thereby subjecting it to considerable movement in wind. This observation seems to be the same in reports from other countries. However, photographs of a

Tree species	*Mean height (m)*	*Range (m)*	*Sample size*	*Reference*
Pine	13	7–19	17	Reinsch & Warncke (1971)
Pine	11.1	2–19	–	Feige (1986)
Deciduous	5.9	1.3–20	343	Feige (1986)
Cherry	0.8	–	–	Zippelius (1972)

Table 24. Published records of nest heights.

Portuguese nest in ash by M. D. England (in Bonham & Bonham 1971) show one placed in a branch at a downward angle. On two occasions GOG fieldworkers have found nests in almost upright branches in poplar, reminiscent of the way Reed Warblers position their nests in reed stems. Another photographed exception appears in Buxton's account of Golden Orioles in his garden in Geneva, where a nest was placed next to an upright bough about one and a half inches (3.75cm) thick in what looks like a cherry tree, with two horizontal angled twigs which met at the point where they joined the upright bough. It looks as if the tree was not very high. The tree must have also been in a protected place as the nest was still intact 12 months later.

Ninety percent of fenland nests are placed in a right-hand fork, the significance of which, if any, is not understood. We can find no reference to this in the literature. Our research has revealed that orientation of the nest from the central trunk is of some importance. Most are placed on a branch between north and south on the eastern side. The remainder are on the western side, but mostly where they are well sheltered by adjacent trees. We can suggest two reasons for this preference: first, the preferred orientation provides protection from wind, as leaves on the rest of any tree would baffle the moving air before it reached the nest; secondly, such a placing would similarly protect the young in any prolonged spell of hot sunshine.

There is no supporting evidence for this finding from other countries. Etienne (pers. comm.) reported that he could find nothing in his Marquenterre study area to suggest significant preference for one orientation over another. Hovorka (1991)

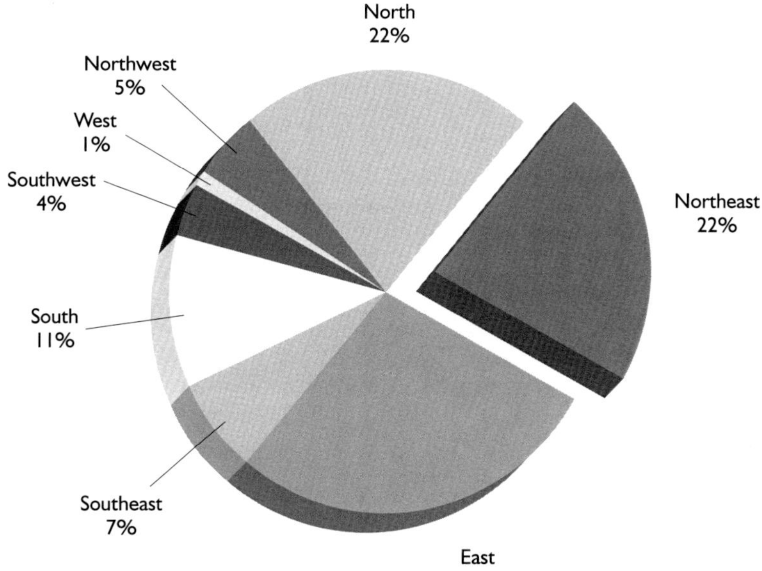

Figure 16. Orientation of oriole nests in poplars in the Fenland Basin study area.

found that in Austria there was an equal distribution, with about 25% in all four quarters. While the predominant wind direction was westerly during the nesting season, winds often came from both a south-easterly and north-easterly direction, causing what Hovorka describes as a Pannonic wind. Under these conditions it will probably make no difference where the nest is placed in any given tree. So it may be that in fenland the prevailing direction of weather has caused orioles to make a definite choice of orientation to provide maximum protection.

NEST-BUILDING

It is the female that actually builds the nest. She is accompanied at all stages by the male, who serenades her with song almost continuously. However, Feige (1986) records one instance of a male collecting material and building in the early stages, and Bezzel *et al.* (1980) include photographs of this. A word of caution is appropriate; when identifying the sex of the bird in the context of unusual behaviour, it should be borne in mind that many very old females can be quite yellow and can be taken for males. The female can travel up to 2km to collect material but most observations in England have been of collection very near the nest. Ray Dowe, a Norfolk gamekeeper who regularly had orioles nesting on this land, watched a female early in the morning stripping bark off a fallen poplar branch and then weaving it into the nest and around branches over a period of at least two hours. During this time, she used only the stripped bark, which was immediately below the nest tree. When the nest was examined after the season, other materials were found to have been used and the nest decorated with feathers both inside and out. PD observed similar behaviour of a female collecting grasses and *Phragmites* reed seed-heads in close proximity to the nest. PM watched a nest in Cambridgeshire built from about the quarter stage to completion over a four-day period. The female incorporated a considerable amount of sheep's wool collected from barbed wire fencing around a sheep's paddock. Another record is of a bird collecting seedheads from ornamental pampas grass in someone's front garden (AB).

The nest itself is of a bowl shape, slung hammock-style in a fork. Some dimensions are given in the Table 25. Weight of nests is in the range 20–30g (Cramp & Simmons 1993).

Authority	*Inner diameter (cm)*	*Outer diameter (cm)*	*Depth (cm)*	*Wall thickness (cm)*
GOG	mean 8	mean 9.5	mean 6	1.3
Feige 1986	mean 8.5	mean 12.5	mean 6	–
Kozlov 1988	range 9.5–10.3	range 8.5–11.0	range 6–8.3	–

Table 25. Dimensions of Golden Oriole nests.

Many people, on being shown examples of nests taken down after use, remark how small they are when judged against the size of the bird. At all times both head and tail of the brooding bird protrude over the edge of the nest, even when the bird is sitting very tight. Twenty-five nests inspected by Reinsch & Warncke (1971) and Feige (1986) showed that the angles of forks selected by the birds ranged from 24 to 65 degrees. When complete, a nest can be very robust, especially if it includes man-made plastic material. Even though placed in such a susceptible position, most are still intact after leaf fall, and a few can even survive well into the following year.

The period of time taken to complete a nest is reckoned by most observers to be between six and 12 days. It is very difficult to find a nest at the 'just-starting' stage, as many are hidden well in the canopy and only become obvious when activity increases. A variety of behaviours is evident from reports by different observers. Several report building to be carried out all day long. Labitte (1951) observed one female that worked for 14 hours on one occasion. Some say females start building operations in short spells of activity, mostly before midday, but activity increases as nest completion nears. Hovorka (1991) managed to find eight different nests that he watched from the very start during a two-year survey. The females concerned all used grasses and created the first weaving of the hammock shape by securing the stems to the branches with spider's web (not saliva as has been reported in some older accounts). Figure 17 (based on Zippelius 1972) shows how the first stages of nest-building are carried out. The third drawing shows the female sitting inside the nest foundation as she secures more strands to the fork.

Once the nest is complete, the female frequently continues to work on it, repairing and titivating the nest and adding to the lining, even when young are present. Replacement nests are rarely recorded, but when they do occur, mostly when the first has been predated at the egg stage, they can be constructed very quickly in about four to five days. No proof of use of a previous year's nest has been obtained, but sometimes the same tree is used. A different branch was mostly used in Britain, but there are reports of the same branch of an oak tree being used for seven consecutive years (Paulussen 1955). Paulussen also quotes a record from another observer of a pair using the same tree for eight consecutive years (tree species not noted). Elsewhere there are assertions that birds have used not only the same tree but even the same branch and same fork for two years running (Rudow 1887, Bau 1907, Reinsch 1958).

From analysis of materials recorded as being used in British nests, we know that the most commonly used natural materials are stripped poplar bark, *Phragmites* reed heads, and grasses. Other natural items recorded are mosses and cocoon silk; also, in late nests, floss from the seed heads of poplar. At a site close to a house and flower garden, birds were seen to collect flowers, mostly yellow flowers of various *Primula* species that were widely available, and to decorate the outside of the nest with them. Man-made materials include red or blue plastic agricultural baler twine, sometimes shredded by the bird into small, thin strips; pieces of plastic sacking torn from discarded sacks; and paper of various origins. At a site where the nest was over a caravan site, females came early in the morning to collect discarded kitchen towel

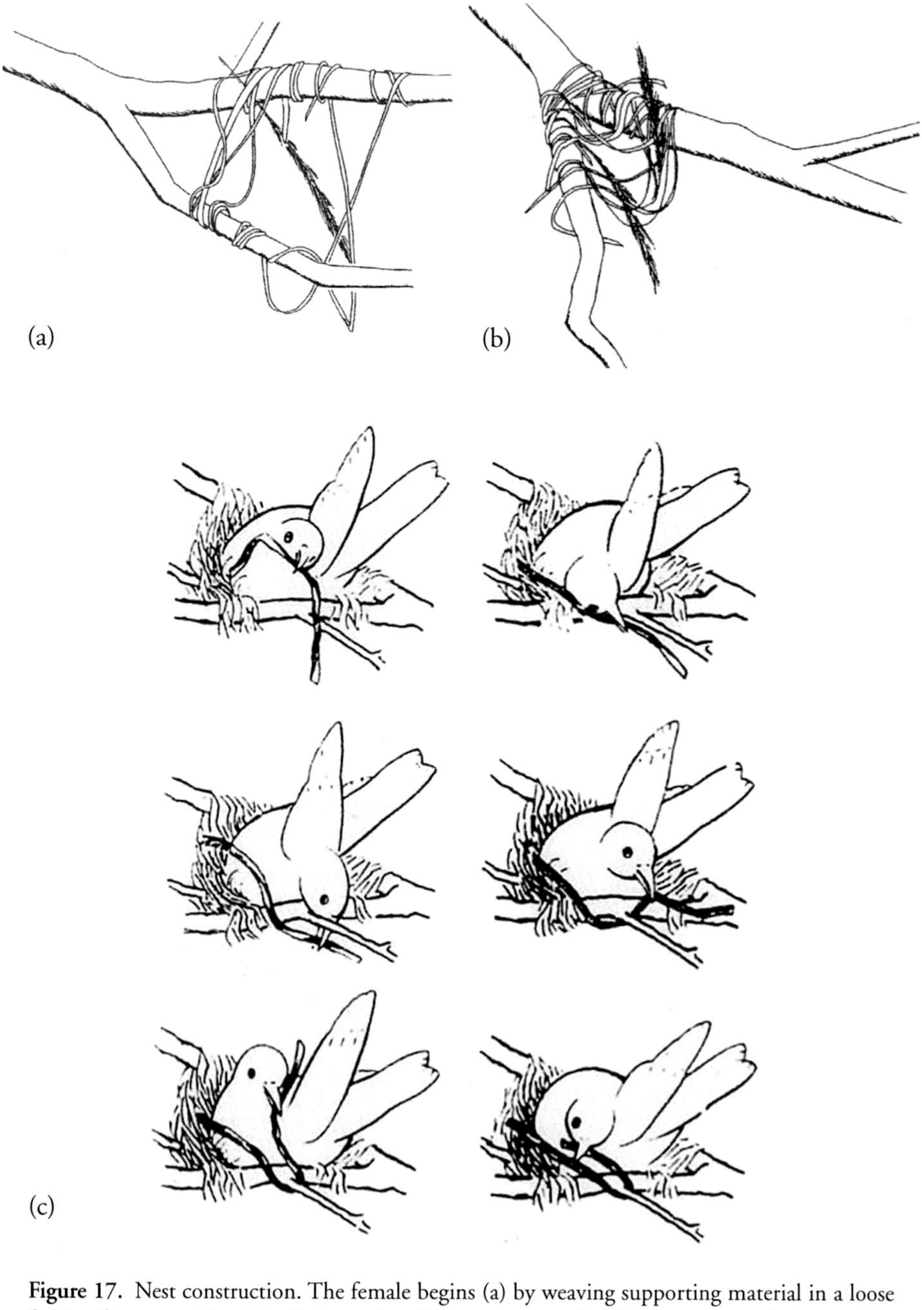

Figure 17. Nest construction. The female begins (a) by weaving supporting material in a loose fashion from one side of the fork to other. She continues to build the nest from the base of the angle of the branch (b), and then sits in the nest, shifting her position as the process of nestbuilding progresses (c). From Zippelius, 1972.

from litter bins. Interesting coloured items that have been recorded as being used in two nests include items of lace put out to dry on bushes after washing (Crette du Palluel 1891). In one case it was a lace cuff with ribbon and the other an embroidered cuff. He also found a nest in which was a crumpled letter, and in yet another a ballot paper and pieces of coloured pictures. Perhaps the most grotesque situation was the finding of a nest containing broken eggs during World War I on the field of battle in the Somme, close to a field ambulance station, constructed from discarded surgical field dressings. These were bound round neatly with string. Included in the nest was paper that was part of a letter addressed to the unit (Coward 1920). There is also another interesting account of a nest found in the aftermath of the Battle of the Somme near Perrone. Congreve (1918) was working there in June and July 1917. The countryside around the town is open and undulating but at that time all the villages, orchards and cultivation were destroyed during the German retreat. There were some small woods still remaining, and at the edge of one of these he found a nest in a slim beech tree about 7m high, under which were old German trenches and dug-outs. It was a puzzle where the fine sheeps' wool came from that was used in the nest's construction. There were no animals in the countryside for many miles around. He found the answer when he came across several scattered woollen mattresses about 100m away, at the entrance of one of the dug-outs. Another nest in a birch tree was made entirely of paper and one or two large chicken feathers. He surmised that it was German paper, maybe used for battle orders. It is sad to think that, as he said he was only interested in birds 'oologically', given the opportunity he would have cut down the nest in order to take the eggs.

Bannerman & Lodge (1953) write of a note received from a Capt. Gilbert, who saw an amusing episode of an oriole collecting items from a scarecrow in open fields in Hungary. Wool gathered from snags on barbed wire has been noted by the GOG and by others, e.g. by Etienne (1988), when he found that the composition of a nest was, on the outer layers, *Phragmites* reed and wool; the middle, cup layer was of wool only; and the inner layer, on which the eggs were actually laid, consisted of mallard feathers. We and several other observers have also found feathers, mostly breast or down. A nest photographed by Valery Moseykin in south Russia contained a very large amount of what looks like dove or pigeon breast and flight feathers, enough to indicate that there may have been a corpse nearby that had been plundered repeatedly.

In a paper written by Csörgy (1911), it was said that nests of orioles were collected to be burnt in Hungary at one time because the smoke was thought to heal bovine mastitis in cattle.

SITE FIDELITY

Evidence gathered in the fenland study area, partly through ringing data and partly through detailed observation, indicates that there is strong fidelity by orioles at the core sites, but less at outlying sites, where breeding tends to be more sporadic.

The data in Appendix 4 show how each of the 37 sites in the study area was occupied during the period 1987–2005, combining recording categories 1–3 (definite breeding, probable breeding and bird/s recorded but no further information). The peaks in Suffolk and Norfolk represent core sites where occupancy was close to 100%. The sites of lesser occupancy are the outlying sites where breeding could only be expected in bumper years.

At Lakenheath, before the bulk of the trees were felled and when the breeding territories were close to each other, site fidelity was hard to assess. Up to 1982 there were 446.4ha of wood, and in 1981, 14 pairs were present, giving a rate of one pair/33.1ha, and this was, of course, in an almost continuous plantation. It is important to note that the birds do not nest in the centres of such large plantations unless there is an open glade formed by some trees having fallen. Most nests are near rides, in the case of Lakenheath some having been as a result of a railway line running through the whole estate. It was easier to come to any conclusions about sites away from there. For example, the most constantly used was in Cambridgeshire (CAM 01), where the ringed male was known to be 11 years old before he no longer returned. The territory was occupied from 1983 to 1997 by at least three different males. Females mated to those males totalled four at least, but there was some overlap between different males. The territory consisted of a 2.5ha balsamifera poplar plantation and two 0.5km double belts of robusta and other hybrid poplars, with a small, three-cornered, open poplar stand of 30 trees at its far end.

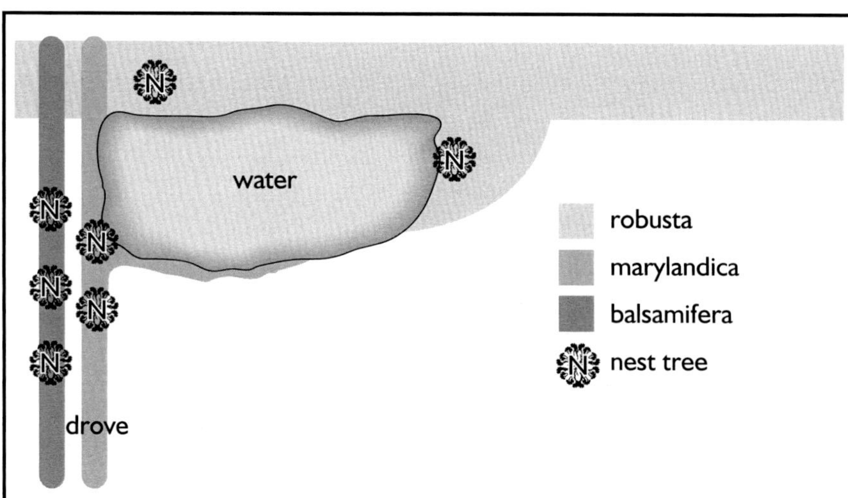

Figure 18. Placement of nests at a Cambridgeshire nest site over a period of seven consecutive years. Birds had used this site before the GOG survey but nest sitings are not known.

At another Cambridgeshire site (Figure 18), which PM first surveyed in 1982, birds nested for seven consecutive years before nesting ceased. Interestingly, local farm staff said the birds had been present at least two years previous to 1982, suggesting a minimum nine-year occupancy. Although it cannot be assumed that the same male was present for the whole period, the female probably was. This is suggested by the fact that her plumage aged through the years from very streaky to very yellow. During the seven years of study at this site, at least 21 young were raised. Normally, one would expect offspring to take over the site after the disappearance of the old birds, but in the following winter, Rooks established a rookery, which almost certainly deterred the orioles from breeding at that site again. If it was the same female for all that time, including the two years before PM studied the site, she would be a candidate for the oldest female in our study area.

A third Cambridgeshire site was occupied for 11 years. We knew that we were dealing with the same female in successive seasons because she had been caught and ringed near the site during the first year of breeding and had been photographed at the nest in a subsequent year. The pair bred almost exclusively in a plantation of 36 mature poplars. They were all hybrid cultivars of mixed types but the nest was always placed in robusta at either end of the block. The only time the pair did not use the old trees was in the first year of occupancy, when the nest was placed at a height of about 3m in a stand of trees about eight to nine years old.

At the other end of the fidelity scale was the case of an isolated west Suffolk site, some 10km from the nearest contemporarily used site in 2002–03. This site was a new one, and in the first year there were two males and one female present. The female may have been a first-year, inexperienced bird as only one young fledged. The following year neither adult male reappeared but a female did, presumably the same one. It was difficult to assess what was occurring until a nest was found fully built some two weeks after the average annual date, but two green birds were present and they were very wary. There had been little or no singing, although contact 'catcalls' had been heard previous to finding the nest and not much later, but by close observation we were able to ascertain that this was a first-year 'green' male mated to a green female, and unless he was observed singing, mostly in a broken and immature fashion, it was difficult to know which was which. They went on to raise at least two young. After the season the nest was inspected and was found to be a good, strongly built example, suggesting that the female was of an age to have had previous experience of nest-building. The fact that the older male from the previous year did not return but was replaced by the first-year male suggests strongly that the young bird might well have been the fledgling from the previous year.

Of the Golden Oriole Group's recoveries, the first was of an adult female recovered 406 days later, having been caught up in wire round a pheasant-rearing pen (not by the ring, we hasten to add). She had nested successfully above the pen in the previous year and no doubt snagged herself whilst gathering food. She had a brood patch. The second was the only adult caught by the Golden Oriole Group ringers. After netting on 17 July 1991, again whilst feeding well-grown fledged

Country	*Total*	*1yr*	*2yrs*	*3yrs*	*4yrs*	*5yrs*	*6yrs*	*>6yrs*	*Comment*
Belgium	1	1							
Holland	2		1		1				
Germany	10	6	2			1	1		
Poland	3	3							
Spain	12	8	2	3					
Switzerland	1							1	7yrs
Greece	2	1		1					
Britain	3	2						1	7 yrs
Lithuania	3								Ages unknown
France+ Tunisia	5		3	1*	1				*Ringed in Channel Is

Table 26. Returns to natal/breeding areas.

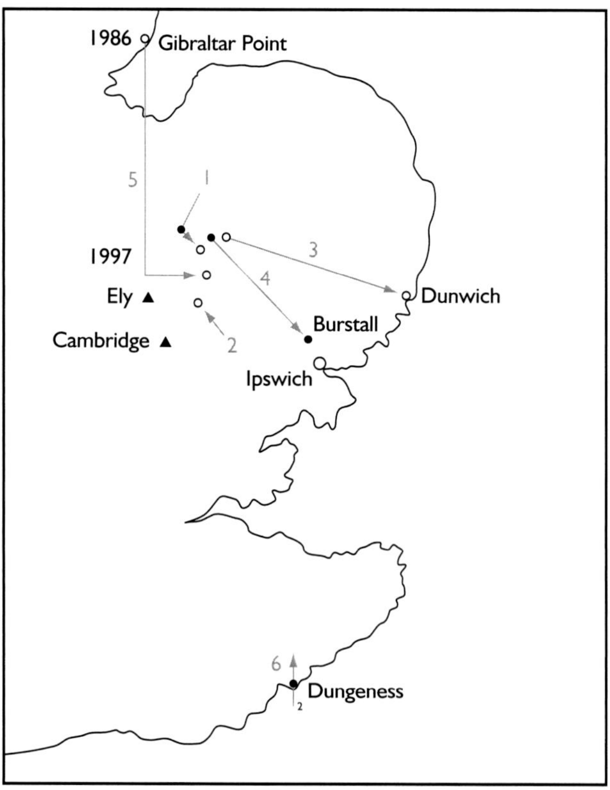

Figure 19. Recoveries of orioles in the Fenland Basin study area (data from the GOG) plus one ringed and one recovered at Dungeness (data for this courtesy of the BTO Ringing Office).

young, the bird was caught at the same site in 1992. The Golden Oriole Group photographers had decided to photograph the nest at that place, and on seeing the hen bird was ringed collated all their pictures and were able to read the numbers. It was found to be the female from the previous year, a neat demonstration of site fidelity.

Returns to natal/breeding areas proved a strong degree of site fidelity as shown by ringing returns from Euring data (Table 26).

ROLE OF NON-BREEDERS DURING THE BREEDING SEASON

In general Golden Orioles are considered to be monogamous, but there is some evidence of pairs having helpers in raising families. Whether a second male who happens to be present is in fact the sire of some of the young in any given nest is not known. It is well known in the case of Dunnocks (Davies 1992) that the presence of a third bird is commonplace. The Golden Oriole Group researchers have frequently observed the presence of a second male at the nest. In one such case, the second male (thought to be a second-year bird) was photographed at the nest, and it was noted that he took no part in tending to the chicks. On the other

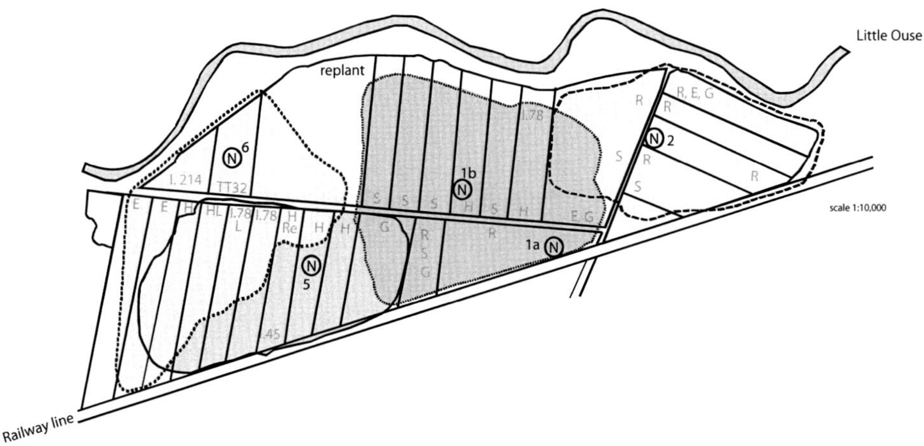

Figure 20. Lakenheath main site in 1987 (i.e. before its purchase by the RSPB) showing the distribution of the main poplar cultivars and the territories of four pairs of breeding orioles. N = nest, with 'pair number'. 1a is pair one's first nest, 1b its second. R = robusta, S = sertotina, H = heidemij, E = eugenii, G = gelrica, Re = rema, L = lons. I.78, TT32, I.214 and I.45 are further cultivars. After Dagley, 1987.

hand, Baumann (*in litt.*) reports that some second males do assist, especially if they are related to the established pair, as proved by colour ringing.

There are many continental reports of nomadic males visiting nest sites, and these reports also say that the resident males chase them away. The available evidence suggests that such attempts to repel the intruding male rarely succeed. In Poland, Ruthke (1951) saw no less than four males feeding and guarding young at one nest. Lavauden (1924) noted a pair plus one extra male actually building a nest. If a pair are split, for instance by the death of a male during the rearing process, a female will sometimes accept the help of another bird, usually an immature (Cramp & Simmons 1993). Etienne (pers. comm.) recorded an interesting variation on this theme when, in 2002, he saw two females attending a nest, although, as far as he could ascertain, only one was the actual parent.

In the dense plantation at Lakenheath, it was seen that, as opposed to fidelity to actual nest sites, the birds' territories overlapped during foraging forays. Dagley (1987) found that the overlapping occurred mostly at the extreme ends of each territory away from the nest placement. He does not say whether birds from one site were present in these shared areas when birds from another were also present; it is likely that they visited the areas at different times so as to avoid clashes. Some pairs had larger territories than others, which is probably a function of the quality of feeding, given that food abundance can vary with poplar clone types. For

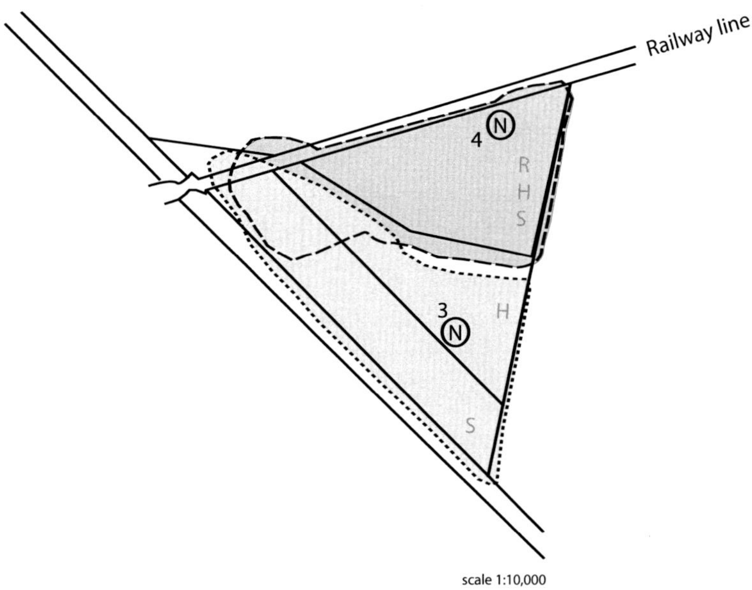

Figure 21. Lakenheath retained triangle in 1987, a little way away from the site shown in Figure 20, with nest sites and territories of two pairs shown. For key to the cultivars see Figure 20. After Dagley, 1987.

example, if the territory included more robusta than serotina, the birds did not need to travel so far because the larger leaves contained more invertebrate food. There was one exception to this rule where pair 6 did come close to pair 5's nest. This occurred, however, where there was a lack of the best types of poplar clone species. Figures 20 and 21 show the extent and proximity of the territories for both the main site and the retained triangle at Lakenheath (Dagley 1987).

CHAPTER EIGHT

Egg-laying and incubation

The eggs of Golden Orioles have a smooth, slightly shiny surface, white or with slightly pink hue, with scattered spots, concentrated dominantly at the top. These spots are dark brown, sometimes with a violet-pink hue.

Average clutch size varies across the oriole's range. From 17 nests in the fenland study area, Milwright (1997) derived a mean clutch size of 3.53 (±0.62).

Researchers in the St Petersburg region of Russia report a usual clutch size of four

Dimensions	*Range (mm)*	*Mean (mm)*
Length	29.3–33.7	31.2
Width	20.7–21.7	21.2
Weight	5.9–6.9	6.3

Table 27. Golden Oriole egg dimensions. From Gavrilov, 1974.

eggs, more rarely three eggs, and no records of clutches of five eggs, and suggest that these figures are typical of other parts of the birds' range in the Russian Federation. In Britain, clutches of five are very rare, and where they do occur, and all five hatch, often one of the nestlings gets pushed out of the nest as the young reach fledgling size. This is probably because the relatively small size of the nest cannot contain all five nestlings, especially where there is excessive wear on the open edge or rim.

EGG-LAYING

After the completion of the nest there is usually a period of one to two days before laying (Reinsch 1958, Feige 1986). In Britain, longer periods have been recorded, sometimes with the pair apparently leaving the nest site altogether. The possibility of their being overlooked, given their secretiveness in dense canopy, cannot be ruled out, however. Periods of up to eight days before laying have been noted, especially during periods of inclement weather. When laying does start, the female stays near to the nest (apart from short feeding forays, when the male stands guard), and often busies herself tiding the nest (Reinsch and Warncke 1971). In the study area, both birds have occasionally been seen to leave the nest area to feed, even when the nest contains eggs.

On the basis of calculated incubation periods from a variety of sources, Milwright (1997) derived first egg-laying dates as shown in Table 28.

INCUBATION PERIOD

From observations in the study area, we have calculated that the period of incubation is normally between 12 and 14 days. In the case of one nest in the fenland study area, the period must have been longer. This nest was found on 31 May, three days before the female started to sit. On 23 June, the nest contained two nestlings of one to two days old and one unhatched egg, which subsequently hatched. Thus, incubation of the first egg must have been not less than 18 days.

Country	*First date*	*Sample size (nests)*
The Netherlands	21 May	42
Central France	23 May	13
Britain	1 June	32
East Germany	2 June	145

Table 28. First egg-laying dates. From Milwright, 1997.

Days (range)	*Region*	*Reference*
12–14	Britain	GOG
13–15	Russia	Korelov 1974
16–17	Germany	Feige 1986
15–18	Palaearctic	Cramp & Simmons 1993

Table 29. Estimates of incubation period.

TIME DEVOTED TO INCUBATION

Observations at a nest in Azerbaijan during the 13 hours of daylight (07:00–20:00 hrs) revealed that about half the time was devoted to incubation, exclusively by the female, distributed as shown in Table 30.

During the period of incubation, the male stays on guard in the vicinity but not necessarily close to the nest. At this time, the whole breeding territory can seem empty of Golden Orioles. The female leaves the nest occasionally to feed nearby and to preen while the male stands guard silently. We have observed exceptions to this rule, when both birds have been seen to leave the nest for short periods, usually only on warm days. Feige (1986) records a male perched above the nest, singing low-key while the female is away and bowing with wings and tail spread. As the eggs hatch, the female removes the eggshells. From this point on, pairs vary in their behaviour and feeding strategy. Most birds take the eggshells a distance from the nest. In the study area, there is only one record of shells being found on the ground below the nest.

HATCHING SEQUENCE, FLEDGING DATE AND PRODUCTIVITY

Milwright (1997) notes that asynchronous hatching appears to be the norm with orioles. He records one nest where there were two nestlings of clearly different sizes.

Time range (hrs)	*Time spent incubating (mins)*
07:00–12:00	120
12:00–16:00	138
16:00–20:00	143

Table 30. Time devoted to incubation on a nest in Azerbaijan. Data from Azerbaijan Academy of Sciences.

Event	*Mean and SD*	*Range*	*Sample size*
Hatching	18 June, 5.5	1 June–1 July	–
Fledging	3 July, –	23 June–15 July	22 nests

Table 31. Hatching and fledging data from the fenland study area.

In six other cases, the nestlings were also of distinctively different sizes, suggesting that hatching of the complete clutch normally takes place over a period of 48–72 hours, i.e. near or above the upper limit of the 12–60 hours observed in central Europe by Feige (1986) and Reinsch & Warnke (1971). One assumes that incubation starts as soon as the first egg is laid, as in Barn Owls, where the difference in size between largest and smallest chick can be considerable.

In order to estimate hatching date of the first chick, given that hatching occurs over two to three days, data from the study area is given in Table 31.

The latest recorded fledging date from the study area, 23 July, involved a late or replacement nest. The Golden Oriole Group has recorded no cases of second broods, but we have had four cases of suspected replacement nests. In two cases, late nests were found one or two trees away from failed nests, and two other nests contained young as late as mid-July.

From 11 broods, it was possible to estimate a mean fledging period of 13.8 days (±0.75), which is much shorter than the published figure of 17 days for continental orioles (Feige 1986, Reinsch & Warncke 1971).

A total of 50 nests in the study area were used to calculate productivity. This involved the counting of nestlings during ringing (29 broods), checking of other live broods (18), and nests containing dead young (3). From these data, a mean brood size of 2.60 was derived, compared with 3.26 from 100 continental nests (Cramp & Simmons 1993).

As to fledging success, Milwright (1997) derived a mean of 2.15 (±0.85), a valuable index for the Golden Oriole Group when calculating the annual minimum and maximum fledging rates. There is some evidence that continental orioles are more productive, as for example in data from Feige (1986) and Malchevski & Pukinski (1983), showing a range of hatching rates from 87.6 to 90.0%, compared with 84.6% from 11 fenland nests that were examined before and after hatching, with presumably a correspondingly higher rate of fledging success.

CAUSES OF NEST FAILURE

The precarious hold that orioles have in Britain means that every adverse circumstance has more serious consequences than would otherwise be the case. The

Desertion caused by human activity (e.g. tree felling)	2
Animal predation (probably corvids)	5
Human predation	2
Weather: effect not specified	7
Weather: nest blown out by gales	4
Weather: nest contents blown out by gales	2
Weather: heavy rain killing young	1
Collapse of poorly built nests	1
Cause unknown	5

Table 32. Causes of nest failure in the fenland study area.

Golden Oriole Group recorded 22 cases of nest failure during the period 1987–96, with causes detailed in Table 32.

Orioles are fiercely aggressive in defence of their nests and can successfully defend them against most predators. It seems that only when there are more than two intruders working cooperatively that successful predation occurs.

The commonest predators in the study area are Carrion Crow, Eurasian Jay and Common Magpie. There are no records of mammalian predators, although orioles have been recorded seeing off Grey Squirrels. Elsewhere in the birds' range, however, predation by Red Squirrels has been noted. In this case several birds from local sites cooperated in seeing off the intruder (Georlette 1939, Labitte 1951).

Of the two cases of human predation, one case involved the removal of the whole nest and chicks, branch and all. One speculates that someone with an aviary fancied a free brood of orioles. It is incredible that there are still egg-collectors in this day and age, but in 1998 a very visible nest at Lakenheath was robbed of its eggs. Investigation by the RSPB and the police led to the apprehension of the culprit, who was fined and imprisoned.

CHAPTER NINE

Nestlings and fledglings

The young are brooded closely for first the five to six days, usually by the female, but other authorities have recorded that the male sometimes takes part (Melde and Melde 1977; Feige 1986). This has not been noted in Britain. At all stages, brooding will take place during periods of inclement weather. Rain, cold and even hot sun will occasion this, even up to fledging. At a Cambridgeshire nest during a period of rain and high wind, the nest, which was situated at about 15m in a single-line poplar shelter belt, was very vulnerable. At this stage, the chicks were about 10 days old. The female could not sit on the birds to brood them, so she gripped a bough on the other side of the nest and crouched over the chicks, but without actually being able to settle down on them. The wind was buffeting the branches of the tree to an extent of at least 3m side to side. She had to do this on at least three recorded occasions over five days, and possibly more. No feeding took place while this occurred. That she was successful in this strategy was confirmed when two to three fledged young were seen in the reeds below the trees. Similar behaviour was noted by PD in 2007, one of the wettest summers on record, when the female straddled the nest, one foot on each side, in order to protect her chicks. Reinsch and Warncke (1971) state that brooding at night stops altogether, three to four days prior to fledging.

CHICK DEVELOPMENT

The chicks are initially blind but their eyes open at approximately six days. At seven days, heads start to appear over the rim of the nest when the parents approach. The chicks grip the bottom of the nest very tenaciously. This makes good sense when one remembers that the nests are often placed at great heights, often out on a flimsy branch and subject to the rigours of strong winds. The combination of the chicks' strong claws and the depth of the cup are clearly intended to guard against the risk of falling out. Even so, there are several records from the fenland study area of stormy weather causing casualties.

The following describes the development of Black-naped Oriole nestlings (Andronova 2003). Given the close relationship of this species to *O. oriolus*, we can assume that the description fits our study species well.

Day 1

The nestlings are practically without embryonic down, except for light yellow down feathers up to 3mm on the eyebrow and the larger upperwing-coverts of the secondaries, and on the tarsi and sacrum. Their skin is thin and of a light yellowish-pink colour. The mouth cavity and tongue are a pale pink, with two dark spots on the upper palate. Bill colour is milky. The tarsi are yellowish-pink; the talons on the toes are of a milky colour. Eyes and ear cavities' holes are closed.

Day 2

The rudiments of primaries and secondaries (as dark points showing through the transparent, thin skin) appear. The eyelids begin to open, the eye continuing to be covered by the nictitating membrane.

1. An eleven-year old ringed male, photographed at its fenland nest. Malcolm Rains.

2. An adult female from Hungary, where Golden Oriole is a common breeding bird. Jari Peltomäki.

3. Fledgling just out of the nest and about to disperse into the foliage. Malcolm Rains.

4. An obvious second-year male, with a green tinge on the mantle and elsewhere. Most noticeable at this stage is the white wing-patch. Malcolm Rains.

6. Adult male Indian Golden Oriole from Karnataka, India. The plumage of this bird differs only slightly different from Eurasian Golden Oriole. Note the deeper black eyestripe extending further towards the nape, slightly more yellow on the tips of the secondaries and the relatively larger yellow wing-patch. Anand Arya.

5. An old female Golden Oriole. This bird is estimated to be 8–10 years old. Females attain yellow plumage as they age but they can be told from males by the greyish black eye stripe and lores (those of males are jet black); also bill is pink (deep red in males). This bird was one of a pair feeding young. Chris Knights.

7. A male African Golden Oriole. In this species the eye stripe goes completely around the eye and the black is more intense, and there is much more yellow on the wings and undertail feathers. Artur Bujanowicz.

8. Orioles feed on both larvae and imagines. Here are some typical prey items from the Fenland area: (a) White Satin larva (Paul Mason); (b) White Satin imago (Malcolm Rains); (c) Herald Moth imago (David Hopkins).

9. A dramatic shot from southern Russia of an oriole with a substantial prey item – a Convolvulus Hawkmoth. Valery Moseykin.

10. Good and bad leaves for orioles. Modern cultivars like beaupre (**a**) are generally larger-leaved and more suitable for orioles to hide their nests from predators. Older cultivars like serotina (**b**) have small leaves that emerge late and provide poor cover for nesting birds. A stand of serotina poplars is also shown (**c**). Paul Mason.

11. Lakenheath RSPB Reserve in May, 2008; a view across the eastern reedbed showing two of the three remaining large poplar plantations used by breeding Golden Orioles (the third is behind the photographer). Others have been planted in recent years but are out of view. Paul Mason.

12. These photographs from the Trebonsko Region, Czech Republic, show how Downy Oaks form a colonnade alongside a large lake (**a**), providing excellent habitat for breeding orioles; and how Sessile Oaks (**b**) are planted so close together that they grow very high (30–40m), making them equally suitable for nesting. Paul Mason.

13. Outside the British Isles, orioles nest in a variety of trees; in this case, in a Cork Oak stand in Spain. Mick Ramsey.

14. South-east Turkey is probably the only place where orioles regularly nest in Oriental Sweet Gum. Geoff Welch.

15. Poplar plantations by a river – typical oriole habitat. Somme, France. Paul Mason.

16. Conifer/birch habitat in Central Siberia. Sergey Gashkov.

17. The egg of the Golden Oriole is bluish-white, lightly speckled with brown. Jeff Baker.

18. This nestling is 7–8 days old, the ideal age for ringing oriole pulli. Jeff Baker.

19. Four nests showing the wide range of materials used by orioles (a–c from Britain, d from southern Russia): (**a**) with plastic, paper and *Phragmites* reed (Malcolm Rains); (**b**) with kitchen roll paper (Malcolm Rains); (**c**) with extensive lining of sheeps' wool (Malcolm Rains); (**d**) with many dove feather quills (Valery Moseykin).

20. A female Golden Oriole incubating her eggs at a fenland site. Malcolm Rains.

21. A fenland oriole feeds an unusual prey item, bumblebees, to its brood of ever-hungry chicks. The bird returns to the nest with three or four bumblebees in its beak, reminiscent of a puffin with a beakfull of sand eels. Chris Knights.

22. In some parts of the range sizeable quantities of mulberry fruit are fed to the young (**a**); soil may be given to the young to help them break down insect food (**b**). Valery Moseykin.

23. This female is perched near the young, using voice and gesture to encourage them to leave the nest. Malcolm Rains.

24. An adult male from southern Russia feeding a fledgling with a Great Green Bush Cricket. Valery Moseykin.

Days 3–4

The 'pin' – tips of feather quills – become visible. Simultaneously, the primary and greater coverts begin to be formed. Feather sheaths appear on the alula and median coverts. The form and colour of the bill start to change to become similar to those in adult birds, i.e. an ivory colour with a light bluish tint. The colour of the tarsi changes simultaneously to a brownish-red, while the spots in the mouth cavity become bluish-violet.

Days 4–5

Simultaneously, feather sheaths appear on the dorsal side of the body between the shoulder blades, on the sacrum and tarsi, on the head around the eyes, and on the ventral parts of the abdominal pterilia (feather tracts). The first signs of the sheath of the rectrices also occur at this time. Changes to the epidermis and mouth cavity begin. The skin acquires a bluish tint, becomes dry and slightly coarse.

Day 6

Sheaths appear near the base of the neck, on the pectoral, collarbone, behind the ear and throat pterilia, and on the crown, the back of the head and the lores. The growth of feathers proceeds centrifugally. Embryonic down is retained on the back (the spinal pterilia). The sheaths on the back and tarsi now begin to emerge open, the feathering showing a yellowish-grey colour.

Day 7

Growth of median and lesser coverts begins. Feather plumes begin to appear on the crown and nape. At the same time the uppertail and undertail-coverts appear, along with marginal coverts on the wings. The number of sheaths on the dorsal and ventral surfaces of the body increase. The nestlings become much more active. Their movements become more coordinated; the muscles of the neck become stronger; they can already hold their heads in a normal position; and they are able to raise themselves up slightly for defecation.

Day 10

The feathers of the remiges begin to emerge. The coordination of movements and feeding reflexes have developed well, and the chicks can now raise themselves up on tiptoe to beg for food.

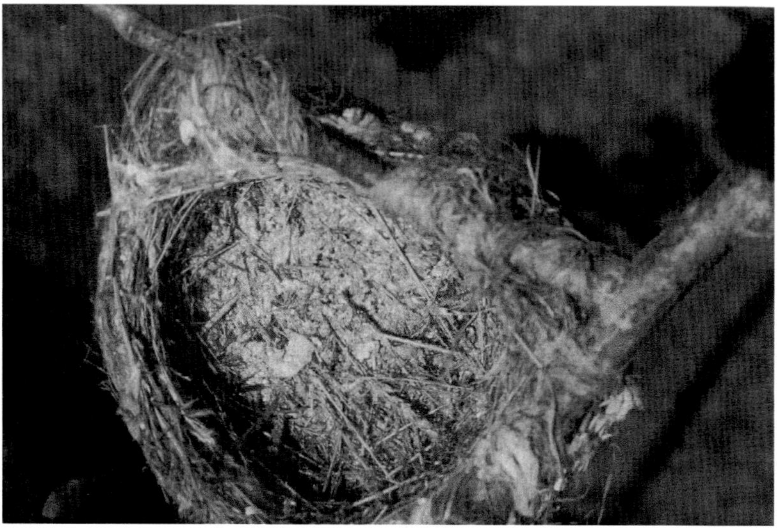

Figure 22. Because nests are often high in trees (25–30m) it is not always possible to be sure that young have fledged successfully, even when adult birds have been present. A retrieved nest containing detritus and feather scales is evidence that young have been reared, and in most cases will have fledged. Paul Mason.

Nest hygiene

In our studies, we have seen parent birds swallow faecal sacs but we have not recorded birds carrying them away, as many other species do. On one occasion a female was seen to take the sac from a chick's cloaca and then to drop it from a nest approximately 12m high. She immediately dropped from her perch, caught the sac before it reached the ground, swallowed it in flight and then proceeded to forage.

FEEDING TECHNIQUES

A variety of feeding techniques has been recorded. The commonest procedure is for the female to feed to the young the items that the foraging male transfers from his beak to hers. The usual procedure is for the male to give a quiet call on approach to the nest, often perching in a nearby tree, where the hen meets him, takes the food and then feeds the young. Uhlenhaut (1999) recorded that one pair effected food transfer at the nest itself. Generally, this pattern does not change for the first two weeks. There are several records of variations on this theme. In some cases, the male will feed the young himself most of the time, with the female very occasionally feeding the odd morsel she might have discovered near the nest as she watched over it. Some other pairs were observed to change their behaviour from about the second week of brooding, up to which time the hen would need to be in attendance in case of the need to cover young in inclement weather. As to the frequency of feeding, in the first half of the fledging period, Reinsch (1958, 1959) noted that the adults returned with food 140 times over a period of 15–16 hours.

After about two weeks, the foraging and feeding becomes a shared occupation, during which time both adults quietly call to the young on approach and young beaks excitedly appear above the nest perimeter. At two sites, we recorded the males both collecting and feeding the chicks 90% of the time, while the female looked on. At one Cambridgeshire site where changeover occurred away from the nest, the female did a small amount of nearby collecting, and on at least one occasion caught a bee, flycatcher-style, by swooping out of the canopy into the open and returning to give it to her young (Chris Knights, pers. comm.). When both parents leave the nest and then one returns to feed the young, the vibration of the parent bird alighting on a nearby branch or twig causes the chicks to raise their heads, gapes open, and to utter begging calls. It seems that once a pair adopts a procedure they tend to stick to it.

Interestingly, whichever sex fed the young, they did not always release the food to the first of the chicks that they offered the food to. While filming at a Cambridgeshire nest, Chris Knights recorded on film a behaviour that may be quite normal but which does not seem to have been documented before. The female approached the nest with a beak full of mixed food and all four nestlings raised their heads, beaks gaping, begging for the meal. The female made as if to put the food into one beak but did not release it. Instead, she made as if to transfer it to the second chick but again did not release the food. She repeated the same procedure with the third chick. When she came to the fourth chick, she actually put the food item into its mouth but then retrieved it. In the end, she fed it to the second chick. All this time, the chicks were uttering their begging calls. Whether this was her normal practice on every occasion was not recorded, but all the young birds fledged successfully and they were all of equal size. It is reasonable to assume that this was the parent's method of ensuring that the food was evenly distributed.

During the first seven days or so from hatching, the chicks are fed mostly by regurgitation. As they grow larger, the young are fed a more varied diet, with whole caterpillars – in our area mostly larvae of the Herald moth. At 10–14 days, large adult moths like Poplar Hawkmoth and Swallow Prominent are fed to young without being de-winged. At a Cambridgeshire nest, PM noted what appeared to be wasps being fed whole to the chicks, apparently without de-stinging. Further observation revealed that the male bird, which had been dropping down to the base of a nearby poplar tree trunk and collecting (even flycatching) the insects, was in fact exploiting the emergence of Hornet Clearwing Moths, which of course needed no de-stinging. Given that these moths look to all intents like large wasps or, as the name implies, hornets, and given that bright yellow and black is a warning of poison or foul taste, one speculates how the oriole has managed not to be fooled by this example of Batesian mimicry; according to Mike Majerus (pers. comm.), the Hornet

Figure 23. Taken at a fenland nest, this photo shows an unusual prey item, a Hornet Clearwing Moth, being brought to the chicks (Malcolm Rains). The inset shows the moths in close-up (David Hopkins).

Moth does not taste foul, despite appearances. Waring (pers. comm.) suggested that it might be only this particular male that had discovered that this moth is an acceptable food item.

Mooiman (1941) noted a parent repeatedly swallowing and regurgitating a beetle until it was softened up enough to be swallowed by a chick. The juice of cherries has been seen by Zippelius (1972) to be dropped into the mouth of each chick before the remaining pulp is fed to one of the young. Valery Moseykin (*in litt.*) photographed mulberries being fed to young at a nest in southern Russia. He also photographed soil being given, something which has not been remarked on elsewhere. With so many juicy and easily digestible Lepidoptera larvae being provided, it would be surprising if any digestive grinding aid was necessary. However, where Moseykin carried out his investigations in 2005, at a nest near the Mynach wetlands, at least 75% of the diet consisted of large Orthoptera such as the Great Green Bush Cricket, prey items with tougher, almost shell-like exoskeletons and wings. One might expect that the digestive system would need a bit of help when dealing with prey such as this. Shrikes and other birds that feed on such insects often regurgitate pellets, but this has not been observed in orioles. Moseykin also noted that at the nest he was studying no larvae at all were fed to the young, which is again very unusual.

MULTIPLE ADULTS

Ruthke (1951) reports up to four males and one female feeding and guarding young. On one occasion, similar behaviour was noted at a fenland nest which was being photographed, when a probable second-year male (judging by its plumage) approached the nest. On three occasions he was seen to look down at the nestlings, but did not feed them, and then flew away. Some ornithologists regard such birds as 'helpers', but others assert that these visitors usually take no part in the breeding process. Feige (1986) called them 'Zigeunerpirole' – gypsy orioles. Sabine Baumann (*in litt.*) confirmed that the most intruders did not act as 'helpers', but she had at least two instances where they did actually contribute to the raising of a family of young orioles that were not their own. She watched second-year males that were the offspring of a certain pair of orioles from a previous year (identified through colour ringing) actually feeding the nestlings. In these two cases it was only after the male parent bird had disappeared that they 'took over', helping the female (their mother) to feed and raise the young. This phenomenon is quite common in birds, and there is a large bulk of literature on helping behaviour. Normally the helpers are closely related; it was first observed in colonial-nesting birds like bee-eaters and some kingfishers.

A curious incident at one of the Golden Oriole Group nest sites is worth recording. We had decided to attempt to trap and ring the parent birds, setting the nets at a safe distance from the nest tree. By this time the fledglings were about seven to 10 days old but not yet at the stage where they would be disturbed by our activity. Nets were set up early in the morning, and as the orioles began to move out of the main trees, we were amazed to see not two but six adults leaving cover. This was a fairly isolated site and no more than the original pair had been noted there before. Where the other four adults came from we do not know. Unfortunately, we were not able to determine whether the additional adults were interacting with the fledglings.

FLEDGING

As the pulli grow, begging calls become louder and can be heard from some distance on still days. As the moment of fledging approaches, there is much flapping of wings over the nest rim. A day or so before fledging actually occurs, the female parent appears to adopt an attitude of encouragement by not feeding the chicks but instead standing on the rim of the nest and peering purposefully round, under and above the nest. Sometimes she stretches her neck and reaches well under the nest, as a sign to the chicks that it is time to leave the nest.

There are records of parents luring young out of the nest with food and calls (Mooiman 1941; Ernholdt 1966). When eventually they do get on to the rim of the nest, the young flap and stretch their wings more and more excitedly. This may continue for a day, with a resultant flattening of the rim edge where it is woven into the supporting twigs. The fledglings eventually launch themselves one at a time from the nest into nearby foliage, constantly uttering a loud 'chip-chip' call. The most developed individual will now flutter from one branch to another farther from the nest, and at this stage, especially in a high nest site, sometimes miss its footing and flutter to low bushes or even to the ground below the tree. At a Cambridgeshire site, DH came across a hapless fledgling on the edge of a ploughed field, where it was very vulnerable to predation. In an attempt to ensure its well-being, DH procured a butterfly net, but before he could make use of it – a tricky business at the best of times – the male arrived and fed the chick on the ground. Shortly thereafter, the fledgling fluttered weakly into a stand of *Phragmites* lining a deepwater ditch. At this point, it was felt that more harm than good would be done by attempting to rescue it. The parent was later observed dropping into the reeds with food on several occasions, so the birds were left to fend for themselves. The family party, including the grounded bird, was later seen safe and sound in the trees. Similarly, PM found all three fledglings from a nest which had evidently missed their footing and ended up perched on twigs growing on the trunk of the nest tree some 5m below the nest. They were left undisturbed and were later seen in a family party, having managed to fly up into the tree canopy.

For the first seven days or so after fledging, the young birds stay hidden in foliage. Sometimes hunger gets the better of them, and their piercing 'chip-chip' calls can be heard, thus giving their position away, to humans at least. On a few occasions they even move into very open positions, where they can be seen quite easily. During this time, the parents will bring food to them, but as the young birds develop, they will start to follow their parents. As time goes on, the young will often split up, with some following the male parent and others the female. At around three to four weeks, family parties have been recorded joining up with others, and at this time even non-breeding birds link up.

DISPERSAL

As the birds get older, dispersal from the natal site proceeds and the diet changes. From late July and early August, fruit becomes a larger part of the diet, and all birds become more and more difficult to find. MR described a sighting where two adults were feeding early plums to a well-grown youngster in an orchard. The behaviour was interesting in that overnight the young bird had found roost cover in a garage and in the morning the older birds were trying to entice it out with the fruit, although at first it was quite reluctant to leave cover. By middle to late August, the young birds forage for themselves and are often deserted by their parents.

CHAPTER TEN

Diet

Across its range, the Golden Oriole's diet is primarily insectivorous during the breeding season until the fledglings have dispersed, after which various kinds of fruit are also taken. In the Golden Oriole Group fen basin study area, numerous nest watches, both physical and photographic, have produced a great deal of information on the diet fed to nestlings. Some analysis of gut contents from dead nestlings has also been undertaken. A wide variety of invertebrate prey items forms the diet of the young in the nest and in their first week or so after fledging. Caterpillars of many species are much fed to young, including the succulent green larvae of the Herald Moth. Amongst other varieties brought in were Poplar Hawkmoth and Eyed Hawkmoth caterpillars, which were fed head first in the manner of Kingfishers feeding scaly fish to their young. Hawkmoth caterpillars have a spike protruding at an upward and backward angle on the last segment of the body, which might theoretically cause swallowing difficulties, but no such problem has been observed. Hairy larvae are fed without any skinning, amongst which the larva of the White Satin Moth is perhaps the most common. This caterpillar is garishly coloured in a mix of black, red and white and is moderately hairy, all of which sends out a warning of unpalatability to many other potential eaters. This moth species is quite numerous in poplars in the fenland study area, but its availability varies from year to year. Later in the fledging period, large moths are fed and it is evident that the oriole's eyesight is keen enough to see through the camouflage of moth species like the prominents, many of which look like extensions of tree bark when settled.

Our knowledge of the diet of orioles in the fenland basin was considerably expanded thanks to a grant from The Environmental Research Fund (TERF), awarded in 1992, to investigate what food items were available to orioles in poplars during the foraging period of June in particular. It was important to try to determine whether the three main poplar varieties differed in their insect species, and how the new varieties compared. It is possible that food supply in nesting trees determined the site preference of East Anglian orioles. To this end, Andrew Foggo from the University of Oxford was engaged to carry out the work (Foggo 1993). He found that some 145 types of Hymenoptera, Diptera, Coleoptera and Lepidoptera were available but this did not mean the birds actually used all of them.

Elsewhere in its range, a wider variety of prey items has been recorded. Etienne has recorded the moderately hairy larvae of the Brown-tail Moth being gleaned from sea buckthorn bushes on the northern French coast. He also noted Lime Hawk, Eyed Hawkmoth and Lappet larvae being collected, as well as Northern Winter Moth larvae from birch trees. On Heligoland, a long-established migration watch point where orioles are frequently recorded on passage, Joschen Dierske reported a great infestation of Buff-tip larvae on sea buckthorn but that no other species of bird was feeding on them (Heligoland Bird Report 2002).

Sabine Baumann has seen similar prey items being taken, as well as many species of Orthoptera, in her breeding study area in Germany, and also on the birds' Zimbabwean wintering grounds, where termites, caught in the air, are a favourite food. In Austria, analysis of faecal sacs by Hovorka (1991) revealed larvae of the Arctiidae family, including Garden Tiger Moth, and various species of ermine and footman moths. The Garden Tiger larva is well known as the 'woolly bear caterpillar' in Britain.

Dementiev & Gladkov (1954) state that the *kundoo* Golden Oriole has been known to take cicadas and their eggs, tree grasshoppers, and various beetles and bugs. Amongst other invertebrates (many not specifically identified), bumblebees formed quite a large proportion, an observation supported by post-mortems on carcasses. These were eaten without any de-stinging being carried out. Literature on the subject mentions wasps being eaten but Golden Oriole Group research cannot confirm this; we did record a confusion species, the Hornet Moth, being fed to young, a prey item not otherwise recorded in the literature.

Various other sources have analysed detail from guts of dead adult birds. An analysis carried out in 1888 by Zarudni of 16 stomachs obtained in Russian forests produced the following results; butterfly larvae 97, shield bugs 25, cicadas 19, grasshoppers 10, beetles 8, asilid flies 3, snake flies 3.

This showed that 75% of the total volume of food in the stomachs was Lepidoptera larvae. An analysis of the stomach of a nestling (Formozov *et al.* 1950) revealed, in addition to the expected prey items, the larvae of Pine Hawkmoth and a species of prominent, a cockchafer and several arachnids. Vietinghoff (1927) believed that cockchafers formed the highest proportion of the diet in Hungary. However, it is clear that orioles will make the most of any plentiful source of food that occurs in their foraging territory, so his assertion may only be relevant to the area where he carried out his research, not necessarily to the country as a whole. An analysis of stomach contents of orioles caught in the Mordovian Nature Reserve revealed two that contained caterpillars of *Bombyx* moths and two that contained caterpillars of the Emperor Moth.

Often large concentrations of orioles can be met with, attracted by large populations of breeding insects. For example, in the Buzukul Woods in Russia (Orenburg region), oak woods with a serious infestation of *Bombyx* caterpillars, oriole numbers were 20 times higher than in neighbouring forests (Shilova 1953). This could also be the case in the Caspian region, where Valery Moseykin photographed a male and female feeding on large grasshoppers, which formed the bulk of the diet fed to large nestling and fledged birds. He also photographed the same male with a Convolvulus Hawkmoth in its beak.

Research in the Ukraine (Petrushenko & Klestov 1991) records orioles collecting prey from magnolia and castor oil plants, from which they gleaned a number of interesting larvae: Goat Moth, a destructive larva that feeds on heartwood; Leopard Moth; a species of eggar moth; and Large Emerald. Some arachnids were also recorded, including orb-web and crab spiders. Crab spiders can change their colour to fit the surroundings and would be difficult to find, all adding to the evidence that orioles must have very keen eyesight. Dragonflies and mayflies were also recorded but there is no indication of the size caught. We may assume that the species referred to were species of damselfly or the smaller species of dragonfly such as the darters, as the larger species of dragonfly such as the hawkers and chasers would be difficult to catch. On the other hand, PM has observed shrikes catching large dragonflies up to the size of Migrant Hawker, so it is not out of the question that orioles could do the same. In the study carried out by Petrushenko & Klestov (1991), a total of 415 different food items were found, which consisted of 53 different species of invertebrate. There were also records of leaves of nettles, which may have been ingested when the larvae of Small Tortoiseshell were being harvested. Petrushenko & Klestov conclude with the interesting observation that oriole food is found in a range of 'levels' of habitat, from the crowns of trees through different layers or tiers to the surface of the ground and in ground vegetation. They also summarizes the size and length of prey items. They note that linear measurements varied quite widely (Table 33).

Item	*Length (mm)*
Yellow garden ants	3.0
Dragonflies, damselflies, maybugs	9.5–30.0
Migratory locusts	70.2

Table 33. Variation in length of prey invertebrates. From Petrushenko & Klestov 1991.

Within this range of 3.0–70.2mm, almost half (48.8%) the items were in the range 10.5–15.0mm. Feige (1986) quotes an aviculture journal, *Die gefiederte Welt*, where it is recorded that orioles in captivity will not take red ants, but will eat moths of the clearwing family (as the Golden Oriole Group found in the wild), which include the ones that look like wasps.

FRUIT

It is obvious that orioles will eat fruit of various sorts on the journey south from their breeding sites and these will include, depending on where they are, cherries, plums, mulberries and any soft fruit they can find. By the time they get to Mediterranean countries like Greece, figs are recorded as also available. Orioles are not a problem for grape growers and olive farmers because the birds leave before the fruit is ripe (M. Dragonetti, *in litt.*). Moseykin and several other observers report that sometimes considerable amounts of mulberry fruit are fed to young. This is also the case in parts of Romania, where the birds nest in mulberry trees to a great extent. Other observers have recorded all varieties of cherry, and in Georgia figs, grapes and pears were eaten. In southern Russia, it is at this time that orioles start to take in much more fruit, although ornithologists have already recorded a certain amount of food of vegetable origin being used, the birds even descending to the ground to eat spoiled apples and pears, in the same way that thrushes exploit this source in orchards in winter. Khokhlov (1991) watched orioles feeding on bird cherry in the last week of May; this fruit had not been recorded as a food source previously. At midday on several red bird cherry trees, two birds, a male and a female, took turns to gather these fruits. They were hardly likely to be feeding young so this must have been for themselves. Around the village where this was recorded (Podlesnoe in Stavropol region), the Golden Oriole is a common breeding species, but this behaviour had not been noted before. In May 1999, despite the abundance of fruit emerging on the bird cherry trees, orioles did not feed on them. At other places in the Stavropol region in May–June, orioles ignored the bird cherry. The speculation is that the single record noted at Podlesnoe involved two birds which had developed an idiosyncratic taste for the fruit. Overall, however, Russian researchers note that food such as the fruits of bird cherry, cherry, currants, hawthorn and others occupy a subordinate place and are found in the orioles' diet mostly in the second half of summer and in autumn.

Once in its African wintering grounds, Golden Orioles feed extensively on figs, in particular *Ficus craterostoma* (Skead 1995). The small fruits of shepherd's trees (Roberts 1928) and Transvaal saffron (Ginn *et al.* 1989) are also gathered, and are swallowed whole.

UNUSUAL PREY ITEMS

Among unusual prey items, there is a record from France of a Great Green Bush Cricket being taken (P. Etienne, *in litt.*). This was in June. In Austria, the analysis of faecal sacs found pieces of eggshell of Spotted Flycatcher and Common Redstart, the eggs probably having been taken from nearby nests (Hovorka 1991). Feige (1986) reports several observers seeing orioles taking eggs and nestlings for food. Earthworms have been recorded on a few occasions: at Portland Bird Observatory, an oriole was seen on two occasions carrying large earthworms (Robertson 1980); there is a record from Tresco, Isles of Scilly, of orioles picking up earthworms (King 1984); and there is a record from Russia of worms found in a dead bird's stomach.

Wassmann (1986) provides a most unusual record of an oriole raiding the larder of a Red-backed Shrike to obtain a mouse and then proceeding to tear it up. It is not known if the oriole managed to separate the indigestible parts from the meat. If it did, it presumably discarded them at the time: there is no record of orioles forming and ejecting pellets. The oriole also took beetles from the same food store. Finally, there is a fascinating record of an Indian Golden Oriole taking a Flying Lizard.

FORAGING TECHNIQUES

Various techniques are employed by orioles to gather their prey. When collecting small prey items to feed to nestlings, foraging is mainly directly from leaves, both topside and underside, and from bark. A considerable amount of hovering over leaves has also been observed. The leaves host all manner of very small larvae, tiny beetle-like items and small, winged flies, fed to the chicks as a mixed grill, so to speak. As the chicks grow, bee species are added to the diet, confirming the oriole's flycatching prowess, as they arrive at the nest with two to four bumblebees in the beak, in the manner of a Puffin with a batch of sand eels. In 1998, birds were seen hovering over and dropping into sugar beet. On investigation, we found large numbers of Silver-Y moth larvae being harvested. That year, there was a large influx of Silver-Ys and also of Painted Lady butterflies all over Europe.

In Africa, where termites are a frequent prey item, the birds catch them in flight and then fly to a perch before consuming them (Brooke 1970). In the north-west of the Stavropol region in southern Russia, where the oriole is common and even at some sites is numerous, Khokhlov (1991) describes an incident where on 23 August 1990, 15 minutes before sunset, two adult orioles (probably a pair) actively hunted above a pond in Podlesnoe settlement, catching insects from the water's surface. The birds half immersed themselves momentarily in the water, repeating the action several times. He said this was the first time he had seen such activity. Similar feeding behaviour has been noted elsewhere (Collar 1975; Grob 1990).

The oriole's habit of coming to ground level to forage is confirmed by an unfortunate incident recorded from Bulgaria, where two orioles and a Bee-eater were discovered in an orchard trapped in the pest plant Green Bristle Grass (R. Scott, pers. comm.; L. Profirov, pers. comm.). As orioles and bee-eaters are not graminivorous, we may assume that they were attracted to the ground by insect prey such as grasshoppers and crickets. Further evidence that orioles are ground feeders is provided by Etienne (1985), who recorded Emperor Moth larvae, gathered from heathland; larvae and chrysalides of Lackey Moth, from sand dunes; and larvae of the Large White butterfly, whose food plant includes a wide range of Cruciferae.

When the orioles arrive in Britain in the spring, it is quite common to see them on the ground on coastal grass fringes, where they are seen to collect various large black hairy caterpillars, which they peel before eating. These larvae are mostly of the Oak Eggar moth, as observed at Portland Bird Observatory (Robertson 1980). A bird spent two days perching on low bushes and dropping on to the caterpillars in the long grass. It then skinned them, holding them by the head and shaking them or banging them against vegetation until the hairy skin was removed (curiously, orioles do not peel hairy caterpillars before feeding them to their nestlings). Ground feeding has been observed on the Isles of Scilly. The late Fred Wardle, keeper to the Dorrian-Smith Abbey Estate, Tresco, usually found orioles in spring as he walked through areas of bracken, sometimes two or three together catching earthworms and other insects (King 1984). Similarly, on St Mary's, a Rosalind Mitchell saw birds feeding on a metalled road. We may assume, therefore, that this practice is commonplace where birds make landfall in a treeless location. In 1984 at Lodmoor and Radipole RSPB reserves near Weymouth in Dorset between 21 April and 3 May, five incoming migrants were found feeding on caterpillars in *Phragmites* reed beds (Dorset Bird Report 1984). These prey items were almost certainly Oak Eggar larvae, as this is the preferred habitat where that species will have overwintered.

FEEDING IN WINTER QUARTERS

Much of the information we have about Golden Orioles in their winter quarters comes from research carried out by Sabine Baumann. In her paper on orioles in Zimbabwe (Baumann 2001), describing her research there, she records that large hairy caterpillars were being taken by all three species of *Oriolus* present in November to February. Sometimes such hairy larvae were in large aggregations. This was in both mopane and miombo habitat, with each species using different sections of any individual tree to find them. Eurasian Golden Orioles used the higher part of the crown, the African Golden Oriole *O. auratus* fed in the middle band of the tree, while the Black-headed Oriole *O. larvatus* confined itself to the lower sections, sometimes taking prey from the trunk, woodpecker-fashion. The two African species would, of course, be feeding young at this time.*

Miombo woodland is found for the most part in Zimbabwe and consists of tall, broadleaved trees, mainly of the genus *Brachystegia*. Mopane refers to woodland that is mostly made up of one species, the turpentine tree, known in the local language as 'mopane'. Trees vary in height from bush-size up to 4m, and they are subject to large caterpillar infestations. The general impression is that Zimbabwean woodland in general is more orchard-like in layout, with trees spaced apart, similar to park-type woodland in Britain. The canopy is less dense in miombo and there are more substantial open areas between the trees, which allows more light to reach the woodland floor (P. Bircham, pers. comm.).

By the time the birds have reached their African wintering grounds they mainly survive on fruit, but not in all cases. Eurasian Golden Orioles were observed in the townships of Somerset West and Robertson near Cape Town, where the birds were attracted to weeping willow trees, which are not thought to provide a great quantity of food for birds generally. However, in the trees in question there were many cicadas, which are locally called Christmas beetles, because this is the time of year when they emerge and fill the air with their shrill calls (they are of course bugs, not beetles). Large numbers of these insects suddenly evacuated the tree noisily. It appears that a male oriole was hopping around the branches trying to catch the insects, and later the bird was seen bashing one against a thick branch before eating it. The eating of cicadas has been recorded in subsequent seasons.

ORIOLES AND PEST CONTROL

A number of the prey items on which orioles feed have pest status, making the oriole a welcome bird in many situations. A good example is the Green Oak Tortrix or European Leafroller. Marco Dragonetti (*in litt.*), who lives in the Grosseto area of Tuscany in Italy, writes: 'At the beginning of May 2005 I saw seven orioles perched on the same oak near my house, at least three others were nearby. Probably a great spread of *Tortrix viridana* larvae (an oak parasite moth) was the cause of these orioles gathering.' An infestation of these larvae can quickly defoliate an oak tree.

*It is not known whether these species share the same practice as Eurasian Golden Oriole parents of feeding hairy larvae to their young unskinned.

One of the most destructive pests is the Gypsy Moth, a well-known defoliator of many deciduous tree species. Many cases are cited from Russia, where the Golden Oriole is considered to be one of the most useful birds of the forest, because its food from arrival in spring and during the breeding period consists mainly of large caterpillars, which rarely form part of the diet of other birds other than corvids and cuckoos. For example, in the stomachs of eight birds shot in the Timashevskie forest belts of the Kuibyshev region, 81% of consumed insects consisted of caterpillars of the Gypsy Moth. Stomachs of orioles, collected by researchers in a Mordovian Nature Reserve, also contained caterpillars of this dangerous pest. Other research revealed that other food collected consisted almost entirely of harmful insects. Formozov *et al.* (1950) describe how, in 1926, a scientist, M. Sharleman, found Gypsy Moth caterpillars in four out of the five stomach contents that he had analysed.

The Russians have good reason to be concerned when they find evidence of Gypsy Moth larvae infestations. The hatching of Gypsy Moth eggs coincides with budding of most hardwood trees. Larvae emerge from egg masses in early spring until mid-May. As with all larvae they develop into adults by going through a series of moults. During the first three stages of development larvae remain in the top branches or crowns of host trees. At this time they would be available to orioles, which, along with magpies and cuckoos, are able to cope with the hairiness. Although the caterpillars are not large, they probably have sufficient hairs to put off smaller insectivorous birds like tits and warblers. The red and black coloration would also be a deterrent. When population numbers are dense, Gypsy Moth larvae feed continuously day and night until the foliage of the host tree is stripped. This is the time when birds would be able to feed on them. The larvae prefer oaks, but will also feed on several hundred different species of trees and shrubs, both hardwood and conifer, thereby making this a pest of serious concern. Fortunately, Gypsy Moth is is no longer a resident British species, although it is still recorded as a migrant (Waring *et al.* 2003). Its last recorded breeding was in the Fens not far from our fenland study area.

Further evidence that the oriole is a useful pest controller concerns the Great Prominent moth. During a mass outbreak of Great Prominent larvae in the forests of Voronezh Nature Reserve, many orioles were observed, in company with other birds, eating concentrations of them (M. N. Kerzina, pers. comm.). Other pest larvae recorded include silk moth spp. and Pine-tree Lappet, while in the forest steppes, the oriole is said to be the main enemy of the larvae of the elm pest *Uropus ulmi*, and of willow silkworms.

Besides caterpillars, orioles quite often control the numbers of such harmful beetles as the Cockchafer. According to Csiki (1904), out of 49 oriole stomachs investigated, 27 contained remains of Cockchafers, together with the larvae of sawyer beetles. A report on research carried out in Ukraine near Kiev and along the Dnister River (Petrushenko & Klestov 1991) lists a number of pest species among the prey items recorded, providing valuable evidence to forestry authorities of the importance of orioles in the extermination of pests of forest plantations and horticultural plants. The report ends: 'It should be stressed that in the food samples there were slugs, locusts, shield bugs, chafers, leaf beetles, weevils, butterflies and moths and larvae of sawflies.'

Orioles as a pest species

Ironically, when orioles change to a fruit diet, they can sometimes achieve the status of pest themselves. In Russia, after leaving the nest, the orioles undertake nomadic movements on the edges of forest and often penetrate into gardens, where they consume cherries and do damage to crops, also pecking out and eating other fruits and berries. They can cause considerable damage to orchard fruit, such as those in the fruit-growing regions of north Caucasia, where many orioles stop during autumn migration. However, because orioles are such a useful pest-control bird in woodland and forests, local policy is that 'only distractive and not exterminating measures should be taken to protect orchard harvests against orioles'. One example of a 'distractive' measure is the planting of berry-bearing shrubs and trees such as mulberry, the fruit of which has little or no commercial value but which is particularly attractive to feeding orioles and starlings.

DRINKING AND BATHING

No instances of orioles drinking have been recorded in the GOG study area. One assumes that the birds get what moisture they need from their diet of insects and, later, of fruit. However, Miller (1985) watched several orioles crossing back and forth over a river near Tours-sur-Marne in France, and apparently drinking by dipping their bills in water whilst in flight. He relates that the birds were all females or juveniles, while a male was seen nearby merely looking on. The action took place near a weir and lasted approximately five minutes before all the birds moved away. He remarked that Barn Swallows were doing the same thing. A similar observation was at the time interpreted as a feeding activity, i.e. the birds picking insects from the surface of the water. It is not easy to determine in either case what exactly the birds were doing. An account by a casual observer (name unknown) describes how he saw a bird fly out from tall poplars and over a river to dip down and take something from the water surface and return to the trees it came from.

Feige (1986) notes that orioles rarely drink from ground puddles but have been seen drinking from tree hollows where water has been trapped. We have had descriptions of a family collectively drinking from a garden pond next to a Cambridgeshire nest site, and no doubt they were also bathing. Dewdrop drinking from leaves has also been seen, while Wassmann (2004) quotes records of orioles taking droplets of water from their own feathers.

Bathing

We have only two unambiguous records of an oriole plunge-bathing. As PM *et al.* approached the base of a tree where a nest of ringable young was located, the female flew off and was seen to settle on a nearby bridge parapet. While ringing was in progress, the bird plunged from the bridge into the deeper section of the water and rose again to the parapet three times, before ruffling her feathers and preening before returning to the nest after the departure of the ringing team. PM saw another example of plunge-bathing by the same bird nine days later. By this time the young were close to fledging, so she presumably judged it safe to leave them unattended. She left the nest and carried out the same bathing and preening routine, after which she proceeded to hunt in the nearby *Phragmites* reeds, and, after capturing a white moth of some sort, returned to the nest to feed one of the chicks before continuing to preen. At another site a female was seen to leave the nest in the morning and evenings to bathe in a nearby drain that was covered in green duckweed. She briefly dipped into the murky water before returning to a nearby tree to preen.

In Norfolk in 1994 on a very hot day, TM watched an adult female fly out of the poplar belt to a small willow on the opposite riverbank. She alighted in the upper branches of this tree, hopped down the lower branches and proceeded to fly back across the river. As she did so, she swooped low, her breast feathers clearly entering the water. She then flew into the poplar belt. She repeated this action three times in as many minutes. We may regard this as an instance of bathing, but the fact that she had panting young in a nearby nest makes one wonder if she was gathering water in her breast feathers, in the manner of sandgrouse, to cool down her overheated offspring. Significantly, the bird was not seen to preen.

There are records of 'belly-flop' bathing from tall trees, sometimes carried out communally involving two to seven birds and taking place in July/August (Stahlbaum 1966; Collar 1975; Grob 1990). Rain-bathing has been seen in August by observers in Hungary. Observations are recorded of birds brushing their wings against wet foliage whilst hanging upside down with tail spread, turning their bodies purposefully against the foliage and then preening (Csörgy 1911; Dorning, 1954; Bezzel *et al.* 1980; Feige 1986).

Interesting observations have also been made on hand-reared or rehabilitated birds. Ann Mason (*in litt.*) describes how orioles were provided with a large bowl of water on a window ledge and both tame and wild birds would then bathe from the edge of the bowl, sometimes just hopping into the water. In 1952, a Miss P. Meade of Croxton in Norfolk described in a letter how she took care of an oriole for 10 days or so in early September. She found the bird, an immature or female, in poor condition in her garden. Its condition was occasioned, she assumed, by very stormy weather at about that time.The bird was evidently starved, and its wing-tips were drooping almost to the ground for the first three days of its stay, indicating that it was unwell. It fed from a dog's bowl, close to the house, eating scraps of rabbit and dog biscuit left by the house pet, and roosted in the dog's kennel for the first two nights. When not feeding it hid in a cabbage patch, moving between places in 'large ungainly hops'. Where it roosted for the remaining period is not known, but it returned to feed until the last day. From the seventh day it 'splashed' in a puddle in the company of Starlings, though whether bathing or drinking was not clear.

CHAPTER ELEVEN

Interspecific relationships

INTERSPECIFIC RIVALRY

PM has made numerous observations, at various Cambridgeshire nests, of orioles defending their nests and young from other species of birds and occasionally squirrels, including mobbing of hovering Kestrels, mostly by two birds, and the chasing off of Carrion Crows flying among trees in dense woodland. On one occasion, a Sparrowhawk was aggressively and noisily pursued by a senior female Golden Oriole. Both she and the intruder uttered shriek calls as she chased the predator some 250m out of the trees and over open land. On the other hand, a pair of Hobbies that nested only 5m distant from a Golden Oriole nest was tolerated for long periods, until, when the young Hobbies were being fed, the orioles did not hesitate to chase the parent birds away. It is known that Magpies often work in pairs when hunting for food. At one nest site, it was clear that the Magpies were getting too close to the orioles' nest for comfort. The Magpies' strategy is for one bird to act as a decoy to divert the attention of their victims while the other sneaks up to the nest and takes eggs or chicks. This pair of Golden Orioles, however, was not to be fooled: the male took on one Magpie and the female the other, and successfully staved off the threat. Similar behaviour was noted where a pair of Jays was progressing through a single tree belt in an open-field situation. As the marauders approached the vicinity of the oriole nest, the parents sprang into action, again tackling the Jays individually.

At another nest, behaviour was noted that has not been recorded elsewhere. The male regularly sat on guard on a prominent open perch in a dead tree while the female was incubating. Often a party of Rooks would fly above the line of the poplar belt in which the nest was situated. When this happened, the male oriole would launch himself like a rocket in their direction and attack them in a manner which reminded the observer of the WWII dogfights between Luftwaffe bombers and British Spitfire fighter planes that he had observed as a boy. The Rooks were very disdainful of this behaviour but whenever the oriole actually struck them they took evasive action. The oriole then returned to its original perch and continued lookout.

On another occasion, as PM was searching for a Golden Oriole nest, the parent birds were very noisy and agitated. Believing himself to be the cause he carried on cautiously, until, on finding the nest, he became aware of a Tawny Owl perched only a few metres above it. He realized then that this was the cause of the furore and retired a little to watch the orioles' reactions to the intruder. In the end their frustration rose to such a peak that they flew into the intruder with considerable force, causing a quite loud 'biff' to be heard. The owl at this point retreated into the wood, no doubt quite used to such treatment. Regardless of this danger being present for some weeks – the owls were breeding not far away – the orioles never lost any young as result of this or any other predator's presence. At a Cambridgeshire site in 1998, DH saw a male parent leave the proximity of its nest during active feeding to fly towards a similar tree belt some 150m distant on the opposite side of an open field in an eastward direction. Suddenly it swerved in mid-journey and flew straight and determinedly in a northerly direction. It soon became clear that the object of the exercise was to harry a Sparrowhawk some 800m away. The hawk was flying in parallel to the oriole's original flight and was not obviously likely to approach the area of the nest. At another nest, PM arrived to find that the eggs had just hatched, and the female was perched on a nearby branch. A party of tits was foraging close by, and two Chaffinches were also present. As these small birds moved along branches and in and out of foliage, the female oriole swooped down and around them. The finches and tits obviously had no designs on the nestlings, but the oriole was taking no chances. She was a young female, so might have been unsure of the small birds' potential as a threat. There is also a record of a male oriole at a Lakenheath nest chasing off a Cuckoo (A.J. Prater, *in litt.*).

Wassmann (1990) recorded orioles attacking thrushes, though only the smaller species below the size of Fieldfare, and then only when the thrushes approached the nest. The logic of this is not clear when much larger birds like Green Woodpecker and Woodpigeon are often attacked. Etienne (1988) also describes seeing Black-headed Gulls being chased over a marsh by a female oriole where both nested in close proximity. Interestingly this same female was, in turn, mobbed by six to eight Barn Swallows. Perhaps the most dramatic and fatal account is by Crette du Palluel (1891). A pair of orioles breeding in an oak with young in the nest was approached by a Common Buzzard, whereupon both adults attacked it. The interloper managed to avoid strikes by the defenders initially and moved to hide deeper in the tree, but the orioles followed it, tearing out feathers from the Buzzard. In final desperation, one of the parents flew at it with some speed and pierced the raptor's throat with its beak. The Buzzard fell to the ground dead.

This defensive antagonism towards other species is probably carried on for some time after either the dispersal of a family party or maybe after all the young of the pair have been predated. For example, in late June 1993, a female was observed at the Ouse Washes RSPB reserve in Cambridgeshire greatly agitated and subsequently mobbing and chasing a Sparrowhawk. The terrain is open wetland and marsh with

Species	*Intruder attacked*	*Flight but no attack*	*Attack from perch*	*No reaction*
Carrion Crow	30	4	4	
Hippolais Warbler	23		18	4
Common Buzzard	12		2	
Magpie	7	(frequent)	3	4
Woodpigeon	6		23	2
Cuckoo	6		1	1
Gt Sp.Woodpecker	5		3	4
Blackbird	3		2	
Jay	4			
Willow Warbler	2		1	2
Blue Tit	2			1
Hawfinch	2			2
Red Kite	2			
Honey Buzzard	2			
Kestrel	2	(frequent)		
Chiffchaff	1			1
Great Tit	1			1
House Sparrow	1		1	
Tree Sparrow	1		1	
Jackdaw	2			
Other small spp.	1			1

Table 34. Intruders and oriole responses. From Uhlenhaut, 1998.

pools and scrapes, and osier beds and low bushes with a few scattered willows. This is definitely not oriole breeding territory and the nearest known breeding pair was at least 15 km distant. The action described continued for several minutes and eventually the birds parted. The oriole flew some half a kilometre away over the open ground with no others of her species present (J. D. Cater & O. Marks, pers. comm.). The date would indicate the likely loss of her family, as experience shows that they would still have been in the vicinity of the nesting area till well into July. The bird obviously still felt very protective.

In Germany, Karl Uhlenhaut (1998) made an intensive study of 84 pairs of breeding orioles over a period of 28 years. He recorded the reaction of the breeding pair to intrusions of other species which came close to, or were in the vicinity of, the nesting area, as summarised in Table 34.

In one case involving a Honey Buzzard, the predator was attacked after it had snatched a chick from the nest. As to Magpies and Kestrels, there were many instances of these birds overflying nests, but the orioles generally did not pursue them. Elsewhere in the breeding range, there are records from Russia of predation by Kestrel, Black Kite and Long-eared Owl.

Threats to the nest from mammals are much rarer. There are a few records of collective defence against squirrels, but little else.

AVOIDANCE OF PREDATORS

In denser arboreal conditions, approaches through the canopy to the nest can be disguised. In shelter belts of single or double lines of trees, birds approaching occupied nests will enter the tree belt up to seven or eight trees away and flit through the others very quickly to the nest. The trouble is that they are not quite so careful on leaving the place: departure is usually from one or two trees away, although it may not always be from the same side of the belt. It seems that if they have seen an observer then they will depart from the opposite side of the belt. However, in more open situations other techniques may evolve showing a remarkable degree of intelligence.

At a Cambridgeshire site where a nest was placed in a group of only 25 trees, surrounded by large, open, arable-cropped fields with the nearest poplar belt being some 1,500m distant, the male adopted a unique strategy when returning with food for the nestlings (see Figure 24). On one side of the 25-tree plantation there was a regularly used public road running parallel to a deep 5m-wide drain where the water level was 3m below ground level. Over the drain at one point was a field entrance bridge. On leaving the nest to collect more food, the male would fly quite openly across the field to the nearest belt of trees, and forage here or in another belt or plantation area 200–250m further on. On returning with a beakful of insects, he would proceed along one or other of the belts until he reached the end near the road and the drain. He then dropped down below ground level, flew along the

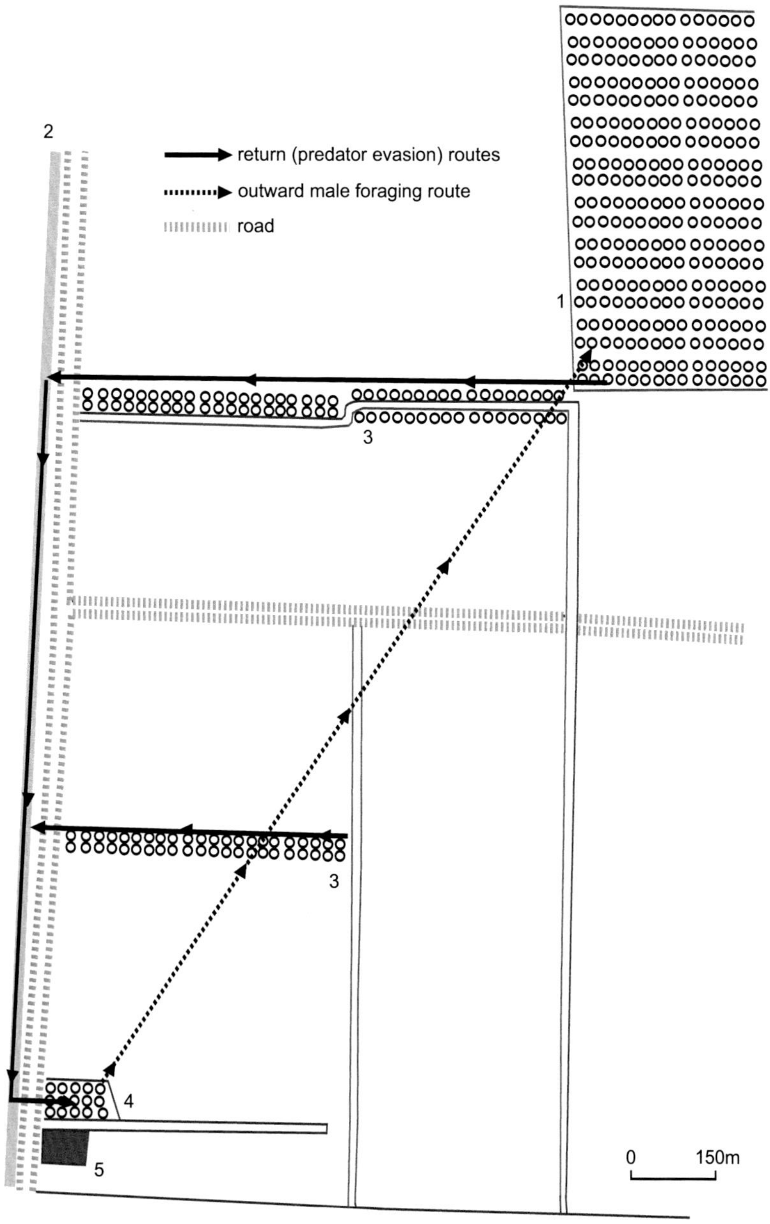

Figure 24. Diagram showing evasive routes taken by a male oriole returning to the nest with food for young at one Cambridgeshire open-fenland site. 1 = large poplar plantation, 2 = deep fen drain, 3 = double poplar belt, 4 = small poplar plantation of c. 30 trees, 5 = house. The routes follow the double poplar belts, then over a road, along the drain until level with the small plantation, then up over the road and back to the nest.

Figure 25. The male at this fenland site would fly below the level of the river bank to return to the small group of poplar trees where its nest was located, having foraged in poplar stands some distance away; these it reached by flying along the belt of trees (Figure 26). These strategies allowed the bird to minimise predation risk during these journeys. Paul Mason.

Figure 26. A belt of poplars, along which the male oriole flew to reach foraging areas. Paul Mason.

drain just above water level, going under the bridge till he was parallel to the poplar stand containing the nest. At this point he would appear out of the drain and fly the short distance over the road to the nest. In other words, after foraging the bird effectively disappeared until it reappeared suddenly close to or at the nest.

On another occasion, birds exhibited similar behaviour, i.e. flying out of sight below ground level. A family party of four birds (a male and three others) were foraging in a small belt of 10-year-old poplars somewhat isolated from the large plantation block where breeding had taken place. The belt was beside a 3m-wide drain where the water level at this time was only about 2m below ground level. The birds were watched by two observers, one on each side of the belt, as they foraged by hovering over and picking items off leaves. Presumably on exhausting the food supply, they were seen to move away, but not over open fields. Instead, they dropped down into the drain and flew below the level of the banks until they arrived at the main body of trees, at which point they emerged and flew up into the canopy. Again one can only assume that it was to ensure that predators were less likely to observe potential prey at a time when the young birds were not yet experienced enough at avoiding danger.

REACTION TO HUMAN PRESENCE

Orioles can get used to the presence of man, especially when there is regular human presence, but dive-bombing can occur even then (von Richter 1949; Malchevski & Pukinski 1983). The extent to which they accept human activities was illustrated when photography was under way in the fenland study site. A mechanical hoist, known as a cherry picker, mounted on a lorry was brought to the base of the target tree a few days before it was planned to start photography. The hoist was gradually raised higher on successive days and the birds never objected. When the time came for the two photographers, Chris Knights and MR, to be in the hide that had been constructed on the hoist platform, they secreted themselves in it at ground level, and then a third person operated the machine until a suitable height had been reached. The third person then had to switch off the lorry engine supplying the hydraulic power, lock the hoist arm into position and depart the scene. After an agreed period, the third party would return and lower the hoist. This cautious procedure was employed at four nests in successive years. In every case, the orioles took absolutely no notice of the intrusion, human or mechanical. Similar indifference to human presence was noted when DM, and later JB, the group's ringers, went up to nests. The reaction of most parent birds was simply to move to nearby trees and emit anxiety calls of varying intensity. Once the ringer had come down the tree again, a watch was kept from a safe distance to ensure that all was well. In every case, the parent birds soon moved back silently. Only on one occasion did a young female feel compelled to dive-bomb the ringer, though so lightly that no injury was sustained.

Figure 27. This cherry-picker has been used by the Golden Oriole Group both for photography at nests and for gathering foliage to allow us to examine insect leaf damage. Proceeding carefully and slowly allows the birds to become habituated to the presence of the machinery. This allows the team access to the canopy without disturbing the birds. Malcolm Rains.

CHAPTER TWELVE

Song and calls

SONG

Listening to a dawn chorus in a fenland poplar plantation with a good mixed understorey is an experience to be treasured. To do so on a day in late May is especially wonderful: a cocktail of songs and calls from Wren, Chaffinch, Blackbird, Song Thrush, Turtle Dove, Garden Warbler, Blackcap, Woodpigeon and Nightingale, punctuated by an insistent Cuckoo or two. But then to have the song of Golden Orioles raining down from the high canopy, giving a tropical atmosphere to it all, is almost like being in heaven. One can liken it to an orchestra playing the opening bars of a concerto, increasing in intensity until the solo flute arrives to play a series of variations on a theme.

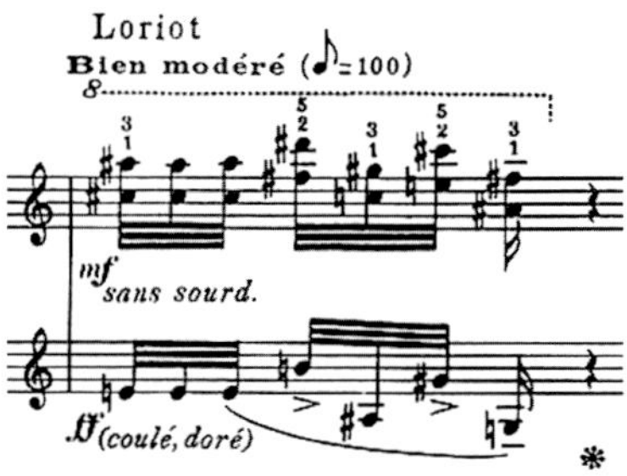

Figure 28. Section of music score as written by Messaien depicting the notes of a Golden Oriole.

The songs of the Common Cuckoo, Nightingale and Woodpigeon are all well-known in music by Handel, Delius, Beethoven, Dvořák *et al.* but the flutelike 'weela weela woolawoo' has, as far as can be ascertained, only been incorporated in Messiaen's *Catalogue des Oiseaux*: 'Le Loriot' (Figure 28). In this avant-garde atonal composition, it may be difficult to recognise the oriole, but in the wild, the song is clear, melodious, loud, fluted, far-reaching and unmistakably beautiful.

In 2007, a work entitled 'The Twitch' was performed on BBC Four Television, composed and played by Peter Gabriel (clarinet) and The Guillemots. It was for a group of traditional and electronic instruments and quite clearly portrayed the song of the Golden Oriole, along with several other British breeding birds.

It is nearly always difficult to interpret bird song and calls in words. Some birds are fortunate enough have onomatopoeic names like the Chiffchaff and the Cuckoo. Others can, at least in English and no doubt in other languages too, have the song described in phrases, like the 'little bit of bread and no cheese' of the Yellowhammer or the 'teacher teacher' of the Great Tit; the Lesser Whitethroat has 'a little bit of bread' but without the cheese. The similarity of the oriole's English and scientific name to its song is fortuitous; the name derives from a Latin word meaning gold. A list of names for the bird in various foreign languages is given in Appendix A5. It will be seen that while some are onomatopeic (e.g. pirol, wywilga), others refer to its coloration (e.g. vlatna, sari, guld) or to its diet (e.g. papa-figos).

Male song

The thrilling, fluty, loud, echoing and sometimes ventriloquial song of the oriole has been described in several ways by different authors with a great amount of permutation on an original theme (see Table 35).

It would seem that most of the lengthy, joined-up songs occur when a male arrives on his site and is trying to attract a mate or at the height of courtship and when there is another male in vicinity. This also happens when the male escorts the female whilst she is gathering nest material and weaving it into the hammock. If another male gets

Country	*Transcription*	*Authority*
Holland	*foh-fluo-fih-fioo, fo-fluh-fuo* *ick-ick-ick ooli-ooli-woo* *weee-ooo-wee, wee-ooo-we-oh*	Bijlsma 1995
Austria	*düdlido, düdlid düdlidi düdl düd dü* *ori eeolio, ee ol i lulo, u li o liloo, loo i lo* *ooliloolilolooo, eeeoooo, oo li li* *hik or hik hik* + the above	Hovorka 1991
Caucasus	*fiu-liu-tiu, fiu-liu*	Sultanov 1988
Germany	*didloh, bio, bio dijo dijo* *didlioh, dihio, didlilio, didlilijoli, didluoh,* *dirojo, dlioh, didjioh,didabio, didljano,* *didijeb, didjudjija, dijauk*	Feige 1986

Table 35. Variations in how the Golden Oriole's call has been described.

too close at this time the song can become almost a shriek at the start and finish and the excitement is plain to hear. When the hen bird is actually incubating, especially at some more isolated sites, the male tends to be quiet. Sometimes, though, when a researcher approaches, the male will often utter a brief 'wee-oooo', sometimes while the intruder is still some way off. After this, silence reigns as the researcher does his checking, but as he departs there is often another 'wee-oooh' to advise the sitting female that the danger has passed. These are all song interpretations at known breeding sites, but of course single, probably first-year, males will become nomadic either from the time they arrive, as in Britain, when they are often encountered well away from known territories, or after they have attempted to annexe established breeding territories unsuccessfully and taken to wandering. They mostly are still green and the song has a kind of 'here I am' sound but is rarely desperate. On making landfall after a sea crossing to British shores, wherever it is, these males are inclined to sing, even when the landscape is most unsuitable, so if they are accompanying others it must be a form of rivalry even then.

Female song

The female has been known, according to several listeners, to give forth a weaker, shorter rendition of the song, and we in the Golden Oriole Group have thought that in all cases when it has been heard it might either have come from a female or it was being uttered by a younger, possibly green, unpaired male entering the territory to see what reaction he would get before proceeding further. Nevertheless Baumann (2000b) has recorded it.

CALLS OF NESTLINGS AND FLEDGLINGS

In the nest the chicks become aware that the parents are approaching the nest, not only from the vibration of their landing on branches nearby but also from the parents' brief, quiet 'widdleoo'. Then heads will rise in eager anticipation and a concentrated 'shweeweeweewee' comes from wide-open gapes, which then fall silent as the parent departs. At most other times the birds are silent unless danger is felt. Once the family need feeding, we have found that there are brief song phrases uttered, but usually just before the male returns with food to the vicinity of the nest but not actually at it. Sometimes it was used to inform his partner to meet him and transfer the food to her to feed the young, but this behaviour was not constant among all pairs. Whenever the nestlings heard it, of course, their heads were raised in excited expectation. For example, in a single-tree line shelter belt around arable fields the male would, on returning with food for the young, enter the belt some seven to eight trees away from the nest and move to around the third or fourth tree from it and utter a brief call, either waiting for the female to come to him or carrying on himself. After the young have fledged, contact between them and the parent birds makes use of a similar call, varying in pitch from fledgling to fledgling. If the recently fledged young are left for some time, a loud insistent 'chip-chip chip-chip' feed-me call is repeated constantly. This usually happens when fledged birds are well hidden in any foliage, but it has also been observed when young are quite prominently perched.

CALLS

Calls, both contact and alarm, are difficult to describe. The call is a cross between a 'cat squawk', as when a cat is angry at being picked up, and a Eurasian Jay's 'kwaaach'. It is almost always only used by the birds when paired up, to act as a contact either for pair assurance when apart or when one or both birds perceive danger. This assurance contact is periodically made by a female while on her nest when she wants to know her mate is still not far away. To reassure her he will either give a short 'weela' or the rasping call by return. When this rasping call is heard, it is reasonably certain that a pair is present, as this call does not appear to be made by unpaired birds. The call is also uttered when there are young in the nest or perhaps just fledged. A quieter form of the call, perhaps at the level of a loudish Eurasian Starling 'steech' or, as one field guide puts it, a hoarse 'veeaak' but a bit more intense, is used for keeping in touch with each other as they go about their business. Malchevski & Pukinski (1983) noted a prolonged twittering subsong, and in the second half of the season, a contact call, 'nrrya'. The latter call is most often made while feeding young, but can also be heard during the period of autumnal nomadic movements and departure. They add: 'In the case of danger, like a threat to nestlings, the Golden Orioles issue a special signal of alarm or threat, which sounds like an extended *yarr.*'

There is one other call, noted by several observers, including GOG fieldworkers, which is a thrilling, falcon-like call of 'kek-kek-kek-kek-kek-kek', with a rising inflection. What brings this on is not really clear, and one has to be sure it is not actually a bird of prey that is uttering it. Danger is not obviously the reason but it could be an expression of excitement at finding a mate, given that, in the fen study area at least, it has only been heard in the pre-nuptial period. It is not clear which sex actually utters it, as it often comes from the deep canopy.

LENGTH AND TIMING OF VOCALISATIONS

Uhlenhaut (1999) undertook intensive studies into the issue of length and timing of vocalisation in the Magdeburg region of Germany. On 17 June, at the time the young were close to fledging, the male was becoming more vocal after a period of relative silence while the hen was incubating the eggs. Uhlenhaut's record, made between 07:44 hrs and 10:01 hrs in the morning, appears in Table 36. The fact that Uhlenhaut's male was not foraging for this period may have indicated that the eggs were close to hatching.

PD, the main GOG fieldworker at the Lakenheath site, has a great deal of experience of oriole vocalisation, and has made some valuable observations. His notes record that birds at Lakenheath most frequently give a two-note call rather than the full, fluty calls that are on commercially available discs. This is interspersed with three-note and four-note calls, but these are much the exception. On several occasions when the bird has been visible, these longer notes have been given when

Time	*Duration (minutes)*	*Vocalisation*
07:44	1	a single song/call
07:47	1	two song phrases
07:50	1	two song phrases
07:51–07:58	8	five song phrases repeated at least 25 times
08:07–08:09	3	five song phrases followed by the catcall
08:09–08:26	18	continuous singing
08:26–08:27	2	one minute of silence
08:27–08:32	6	another series of 23 calls/song phrases
08:32–08:33	2	one minute of silence
08:33–10:01	18	continual singing, interspersed with 10 one-minute pauses

Table 36. Duration of vocalisation on 17 June. From Uhlenhaut 1999.

the bird has moved from one perch to another. The males move around their territory calling and singing from different trees.

On 2 May 2005, PD recorded that a male oriole had three bursts of song and calls between 06:08 hrs and 06:30 hrs.

The total number of phrases for each minute in which song was given are detailed in Figure 29; the number of notes in each song phrase varied from two to four (Figure 30). Similar counting of song phrases was carried out for a period of time from 07:15 hrs until 07:34 hrs on 22 May 2005. The results are shown in Table 37. The bird was then totally quiet until at least 08:15. Similar notes were taken on 14 May 2000 (Table 38).

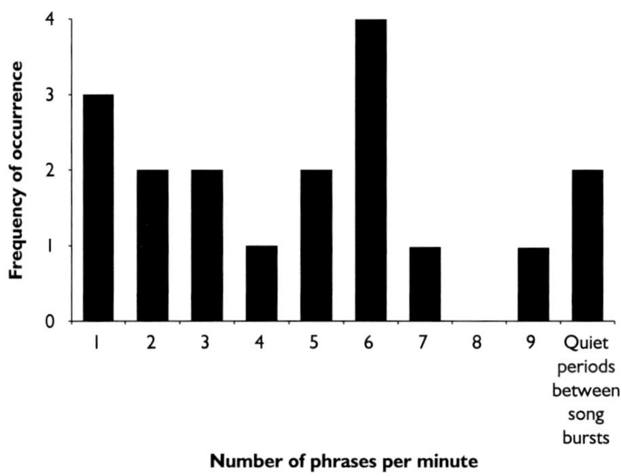

Figure 29. Number of phrases per minute.

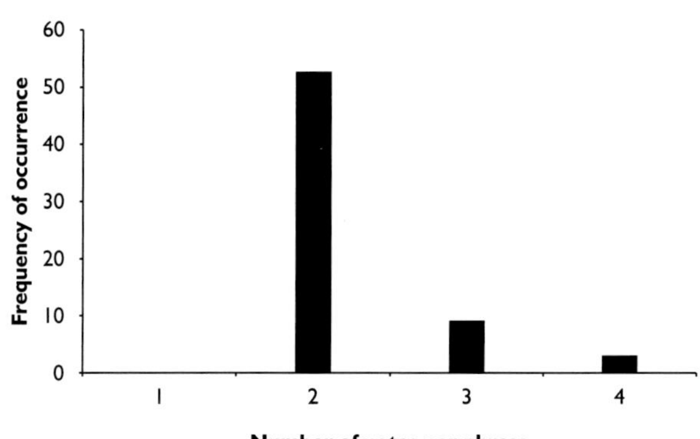

Figure 30. Number of notes per phrase.

One-minute periods	*Total no. of of phrases given*	*No. of two-note phrases*	*No. of three-note phrases*	*No. of four-note phrases*	*No. of five-note phrases*
07:15	7	4	2	1	
07:17	8	8			
07:20	6	5	1		
07:21	5	4	1		
07:22	6	6			
07:23	7	7			
07:28	6	4	1	1	
07:29	6	6			
07:30	9	6	1	2	
07:31	10	8	1	1	
07:32	8	5		3	
07:33	5	4		1	
07:34	3	2		1	
07:35	3	3			
Total	89	72	7	10	

Table 37. Song characteristics from 07:15 to 07:35 on 22 May 2005.

One-minute periods	*Total no. of of phrases given*	*No. of two-note phrases*	*No. of three-note phrases*	*No. of four-note phrases*	*No. of five-note phrases*
05:50	7	5		2	
05:51	9	4		4	1
05:52	11	4	1	6	
05:53	10	6		4	
05:54	11	6	2	3	
05:55	7	2	2	1	2 (of 6)
05:56	10	4	3	2	1
05:57	7	2	1	1	3
Total	72	33	9	23	7

Table 38. Song characteristics from 05:50 to 05:57 on 14 May 2000.

Date	*Sunrise*	*Song first noted*	*Weather*
29/5/05	04:46	04:25	8°C, no wind
13/5/00	05:08	04:30	8/8 cloud cover, still, warm
14/5/00	05:09	05:05	Mist, visibility down to 40m

Table 39. Weather and time of first song.

The first song of the day

Observations made by PD (Table 39) suggest that weather conditions, including temperature and wind, must determine, at least in part, the timing of morning song.

Start time in relation to sunrise varies considerably, the fluctuations undoubtedly related to two main factors: weather conditions and the stage of the breeding cycle. The original data show that between 5 and 18 June, the song was not heard early, if at all, and this would coincide with the period of females brooding, when the male goes quiet. From 18 June onwards, after the chicks have hatched, the male recommences his singing. Singing can then go on for some time, at any time, till about the middle of August. For example, in the St Petersburg region in northern European Russia, in some years the song of the Golden Oriole can be heard from 12 May (earliest date, 1968) to 22 August (latest date, 1981) Thus, the Golden Oriole sings during almost the whole period of its presence in the breeding region.

SONAGRAM RECORDINGS

Interpretation by sonagram is the principal way of describing accurately in print the sound of a song, and several are needed to show variation from bird to bird or region to region.

The sonograms shown on the next four pages were recorded in an oak wood at Poderone, Toscana, in central Italy, 9 May 2004 at 19:30 hrs (Dragonetti).

Adult male song

The interval between each song phrase is shortened to two seconds or less. The mean time interval between phrases is 7.0 seconds, $\pm$ 1.99.

Sonogram 1

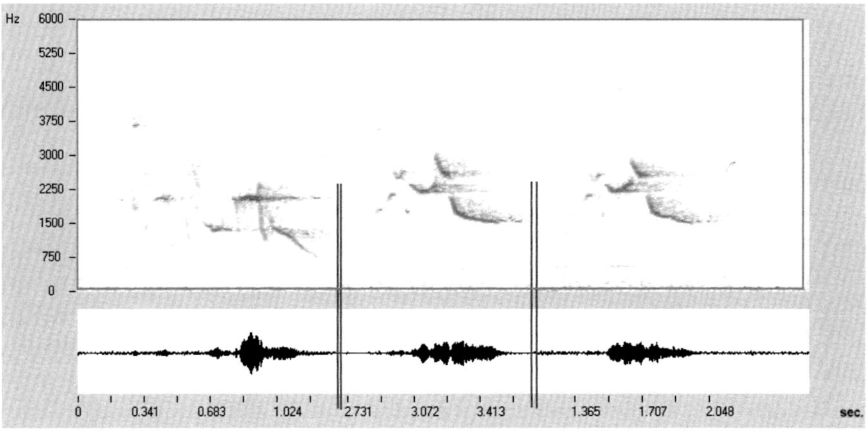

oo oriOOlioo o o oriooloo o ooliooo

Sonogram 2

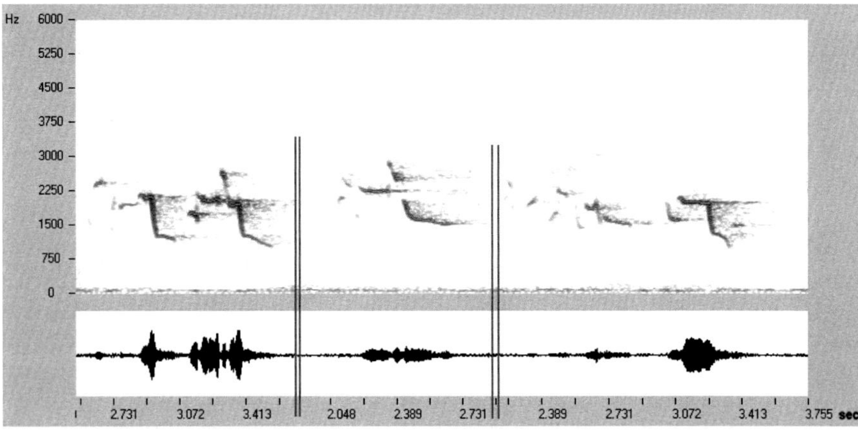

ooOOliOOOo olioooo o o oli oo oOOloo

KEK-KEK-KEK CALL

At the same site, 17 June 2004 at 23:00 hrs, Dragonetti also succeeded in recording the loud 'kek kek kek' call, reminiscent of a Kestrel's contact call. Two birds were perched in an oak wood (probably a pair near the nest).

After a long song duet (whistling song plus full song), one bird (probably the male) uttered many falcon calls.

Sonogram 3

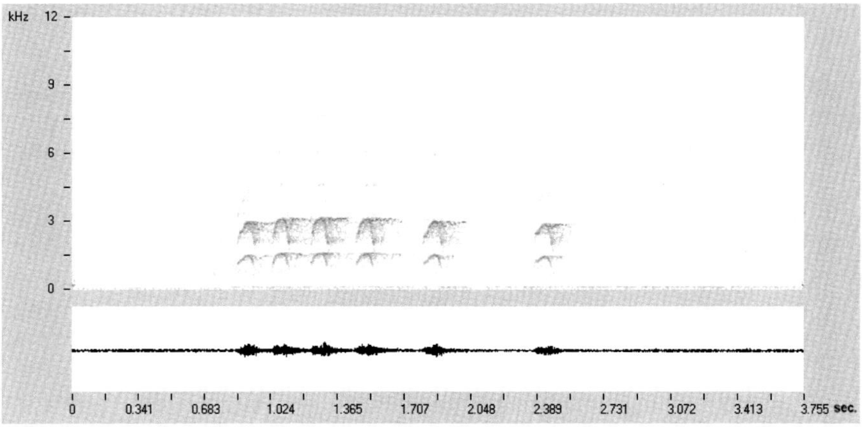

kek *ke-ek* *kek*

Sonogram 4

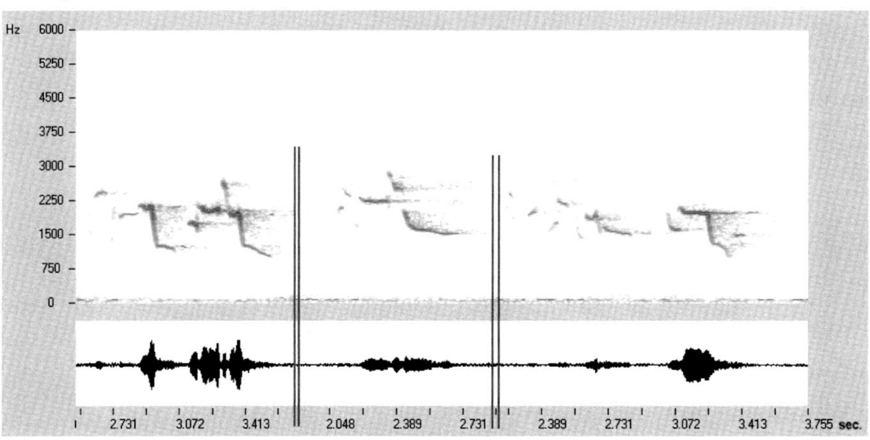

kek ke-ek (intense)

ke-ek (soft)

ke kek

Sonogram 5

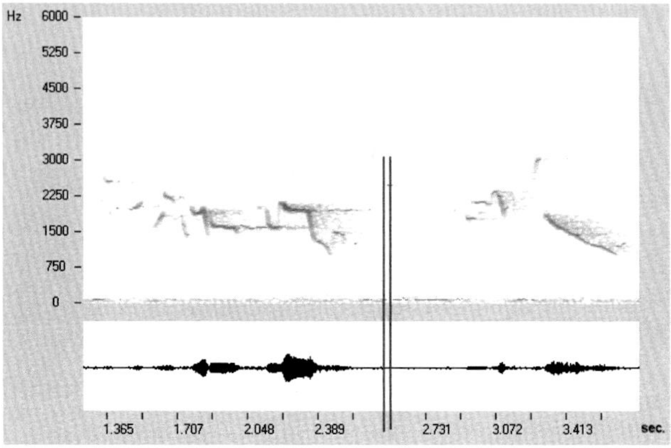

kek (soft) *kek* (loud) *ki ke ik* (fading)

In the *Handbook to the Birds of the World*, this call is said to be made by a female in answer to the song of the male, but neither we nor Dragonetti can confirm this. In both cases the sounds came from somewhere well hidden in the canopy. Sabine Baumann (*in litt.*) said she heard them mostly in an aggressive context from both partners, for example against Cuckoos, Crows or other Golden Orioles.

'Cat squawk' call

At the same site, 16 June 2004 at 07:00 hrs; the 'cat squawk' rasping call referred to is an excited, aggressive alarm call, probably given by both birds of the pair.

Sonogram 6

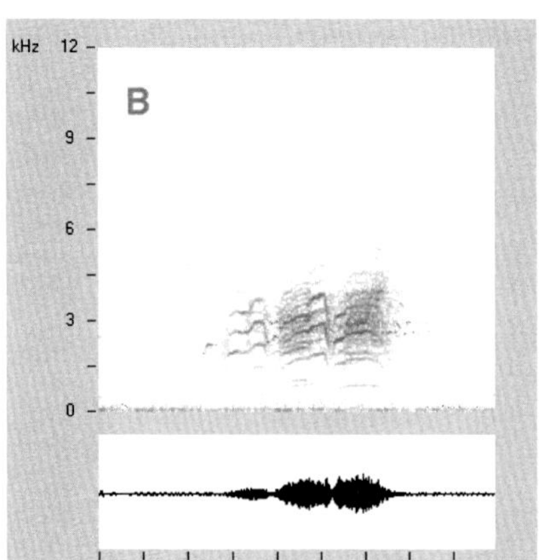

Skaaaaaaaark

Sonogram 7

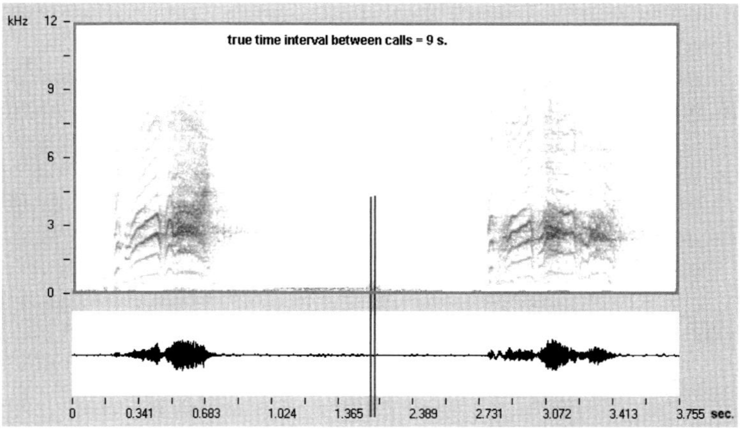

shhH aarrk *shhhhaaarshark*

THE SIGNIFICANCE OF INDIVIDUAL AND REGIONAL VARIATIONS

Researchers wanted to discover more about male Great Bitterns at their breeding sites. Given that Bitterns breed in vast, dense reedbeds, it is impossible to learn a

Sonogram 8

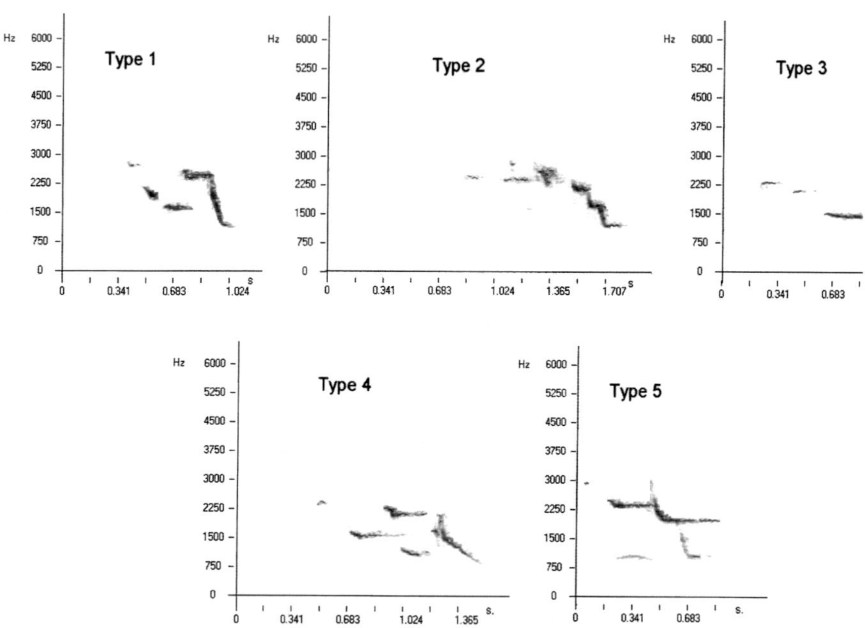

great deal by simple observation. The scientists found that they were able to identify each male by the sound constituents that made up his booming call. Orioles present a similar problem, in that they spend much of their time largely unseen in thick tree canopy. Similar research to that done with Bitterns could do much to support what anecdotal evidence is currently available. According to Baumann (2000b), the analysis of sonagrams will only be helpful if you analyse the total repertoire of each bird.

For practical reasons, no systematic recording was carried out by the Golden Oriole Group, but a 23-minute recording of one male was made in May 1995 by photographer Chris Knights. The recording was sent to Marco Dragonetti, who transferred the original tape to digital CD and split it into five tracks, as shown in Sonagram 8

Total song types analysed = 98
Recording time analysed = about 23 minutes
Mean interval between songs (overall period) was 12.1 seconds (n = 97).
Mean interval between songs (during song bouts = continuous singing) = 7.4 seconds (n = 75).
Number of elements per song = five for all types except Type 1, which has four elements

Track 1 contained no oriole song so is disregarded. The other tracks were almost continuous, except for between 2 and 3, when a short interval occurred. There are a total of 103 song phrases and one squawk call. A total of 98 songs was analysed, leaving aside some four or five of the less distinct ones, where the spectrograms were unreadable. There are five song types. It can be seen that this bird has only a small repertoire, suggesting that it might have been a young male bird; with only five different song types, it has a long way to go before reaching the range of some German birds, which can have from 28 to 46 song types (Baumann 2000b). Further details of Chris Knights' recording are given in Appendix 6.

Sabine Baumann's study (Baumann 2000b), carried out over a four-year period in Schleswig-Holstein and Lower Saxony in Germany, is the most thorough ever done into Golden Oriole song and involved 50 individually recognisable birds, including 24 radio-tagged birds. Nearly 10,000 songs were analysed, and this large number allowed for some important conclusions to be drawn about Golden Oriole song repertoire. Singing behaviour of non-breeding males in the vicinity of breeding pairs, mainly likely to be green first-year, yellow second-year or unpaired older males, was not analysed.

The database of the study was of 50 colour-ringed males, 12 pairs with transmitters and sonagraphic analysis of 9,853 songs. She found that each bird had a few individual song types but the repertoire of any given male may be as great 50 song types or more. There are special song types, for example, that a male uses in communication with its female (who can also be singing by herself nearby) and special song types in contact with neighbours. About five to 10 song types are

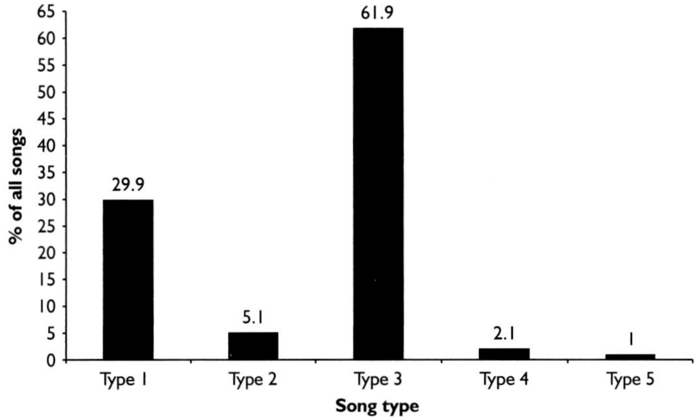

Figure 31. Song-type sequence used by one male in a Norfolk poplar wood site (the same bird as in Figure 32).

specific to one male and exclusively sung only by one bird. There are, it seems, no special song phrases that are unique to only one individual bird, but there are song variations, meaning that there will be characteristic song types in common in different populations. If one is very familiar with a given ringed male one should be able to identify it on song. Moreover, birds in one area may have a 'dialect', i.e. a group of song types that will differentiate them from birds in another area. Her analysis covered the following:

- Description of the repertoire
- Qualitative description of newly found aspects
- Sonagraphic description of the twittering, or subsong, of females
- Sonagraphic description of the song of a hand-reared, first-year male

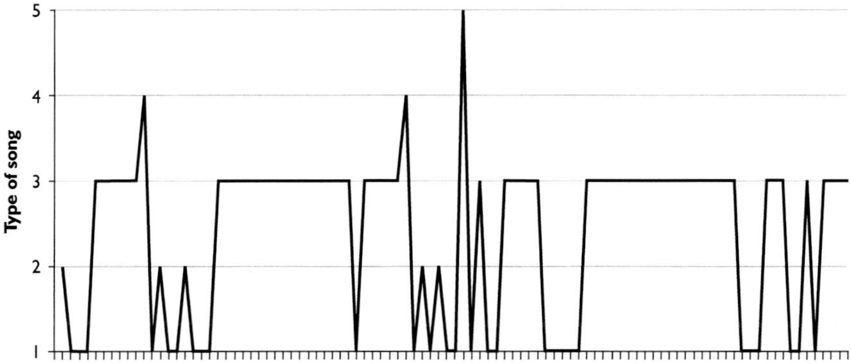

Figure 32. Song repertoire of one male in a Norfolk poplar wood.

Bird	*Number of song types*	*Exclusivity*
Males	35	14 used only by the male
Female	26	shared with male
Helpers	29	shared with male

Table 40. Song type exclusivity.

- Quantitative description of repertoires of 10 pairs and six helpers and test of individual characters
- Dialects in the song of male orioles.
- Habitat use relative to singing in both males and females.

In respect of one breeding pair with attendant 'helpers', she noted that all song types used by the female and the 'helpers' were in the male's repertoire, but that he had additional song types of his own.

Analysis by Baumann of the repertoire of another breeding pair revealed that of 40 song types (calls, subsong and song phrases), none were used by the female only, 14 were used by the male only, and 26 were used in common by both birds (Figure 33).

Songs of the sexes differ in 11 of 19 acoustic parameters studied. Female songs are shorter but some of the frequency ranges higher. In addition to differences at the repertoire level in song-type composition, the songs of individuals also differ at song-type level in acoustic parameters.

The analysis of individual parameters of song types show a variance of 58.9%, caused by two factors:

- 32% are described by the variables of element duration, frequency maximum and frequency minimum.
- 26.9% of variance is described by the parameters total frequency spectrum and frequency spectrum of elements.

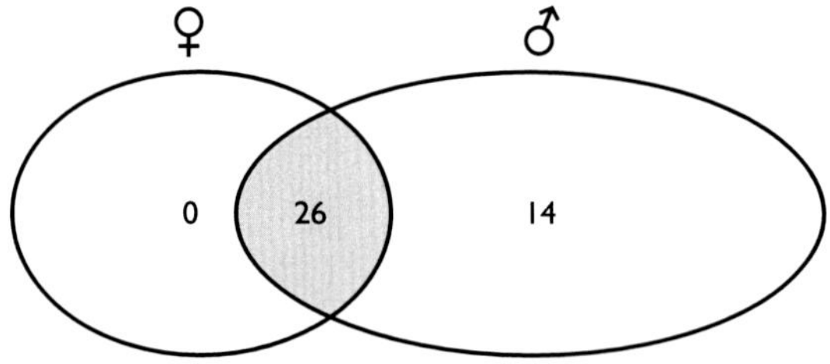

Figure 33. Song phrases used by a breeding pair at a German site; 26 were used by both birds, 14 only by the male. The female had no unique calls. After Baumann, 2000b.

Sex-specific song-parameters are defined as follows:

- 34.3% includes all parameters of element and song duration, as well as frequency spectrum of song and all single elements
- 21.3% includes the variables of frequency minimum and frequency of element 2.

TERRITORIES AND SONG REPERTOIRE

The analysis of radio-tracking data show that there are two kinds of Golden Oriole territory. Firstly, there is a close area around the nest (which we call nest territory), which is used exclusively by the pair and sometimes helpers at the nest. Secondly, there is a much larger territory marked by song-posts (called singing territory), in which 'well-known' individuals like neighbours may be tolerated (e.g. during feeding) while 'foreign' orioles are not. There are different song types used in these two kinds of territory. Each male has exclusive song types used from song-posts in the singing territory and others used only in the nest territory, approximately 10% of the total repertoire in both cases.

Results of analysis of repertoires in three cases of two neighbouring males found that neighbouring males have a pool of common song types found in both repertoires. In territorial encounters with neighbours these are matching repertoires. During counter-singing from song-posts an answering male does not use the song types uttered by his opponent but sings other songs from the common pool of song types. Playback experiments show that these males may be able to discriminate between their close neighbours and others venturing nomadically into established breeding territories.

The singing behaviour of Golden Orioles is multifaceted. Both sexes have large song repertoires, which can be sex specific with individual song characteristics; there are also dialects at the repertoire level and males show specialised use of song types. Female song may show a hint of their tropical origin, because in tropical birds females are often singing or duetting. In the Eurasian Golden Oriole females have only been observed singing in the nesting area: flute-like song types only in the presence of males or during encounters with adult interlopers at breeding sites. Female song may therefore have a social function, for example strengthening pair bonds with partners. This is typical for situations when breeding conditions are not predictable or short-timed. Oriole partners often duet in the nesting area, using song types common to each other. Size of song repertoire may be important in respect of sexual selection. In the Eurasian Golden Oriole males arrive in the breeding grounds several days before females and immature birds. Males with a more complex repertoire may have a greater success in attracting a mate, because ownership of a large collection of songs may be a sign of male fitness. Individual and sex-specific song parameters may allow recognition of partners, neighbours and helpers. This may allow successful mate guarding and identification of outsiders, in turn leading to stable neighbourhoods.

Unpublished data on hand-reared orioles of both sexes show that song learning proceeds over a period of at least of several months. In the field, tutors must be parents, and neighbours and young birds may learn the dialect of the local sub-population. This may influence dispersal and social adaptation in subsequent years. On arrival on wintering grounds some singing has been noted as the birds travel around in groups foraging. All these various sounds emitted by our subject bird add up to it being an exciting species to study.

A NOTE ON DIALECTS

Observations made by PD of oriole vocalisation in several overseas locations lend credence to this notion of what one might call dialectal variations. At Lotan in southern Israel he heard a passage bird on 13 April 2006, which gave 'the full flute-like range of calls'. PD provided a record from Poland of a family party moving through a forest in a loose flock: 'Call notes were much more clipped, written as *weweoo* with the first syllable very short and abbreviated compared with Lakenheath birds. Also gave a quick *ree ree ree* note followed by standard two-note call.' In Slovakia, PD heard a juvenile repeating a two-syllable call: 'the first note not dissimilar to a catcall, but clearer and less slurred, not heard at Lakenheath'. Another uncharacteristic call was noted in Hungary, where PD observed a family in lines of willows and aspen around sugar settling ponds. He recorded 'one plain, cat call. . . and a persistent 3-note call eh-eh-eh, quite unlike what is heard at Lakenheath'.

VOCALISATION IN OTHER *ORIOLUS* SPECIES

INDIAN GOLDEN ORIOLE

The song of the Indian Golden Oriole is generally similar to *O. oriolus*, but, to quote HBW, 'more complex, with shorter, mellower strophes and narrower range of pitch: fluty melodious "pee-lo" or "pee-lo-lo", "who-he-heer" or "weela whee-oh"; most commonly a harsh "chee-ah" or "kaach"; also a nasal hiss "ka-a-lee-aa" or very loud, nasal, upslurred miaowing that starts clear but ends screechy and harsh; sometimes a rapid "tur-tur-tur-tur"; may give a sweet, prolonged undertone song.' Gavrilov (1974) described his experience of the bird in Khazakhstan thus:

> Arriving at Almaty, males begin to sing intensively at the outset. Singing is of two types. Firstly a set of quiet whistles and twittering sounds and at this stage is not loud. The second type is a melodious, loud, flute whistle. As with the nominates, approximately one month after arrival, by the middle of June, intensity of singing begins to lower. The Golden Oriole song is a flutelike whistle, which can be heard at the great distance. It can be translated as 'fiiu-lyu-iiu'.

Gavrilov (1974) writes in turn that 'it was noticeable that in the case of the Indian Oriole (*kundoo*), in contrast to the Eurasian species, the whistle is more melodious and smooth, consisting as if of more separated syllables.'

African Golden Oriole

'Weelo-yoo-tyoowo', 'weelo-lo-weeluweeyu'. Displaying birds utter a longer, drawn-out 'du-wee-weeeyer'. Aggression and alarm calls are a harsh 'rrraaaah' or a cat-like, mewing 'myaaaeeeh'.

African Black-headed Oriole

A singular, short 'keeuu', falling in pitch and repeated a number of times. Its calls consist of a bubbly 'lup-lop, iddliop' and the typical oriole harsh 'waar' contact and alarm call. Playback of all three of these oriole songs to birds in Zimbabwe revealed that Eurasian Golden Orioles reacted to African Golden Orioles' song and vice versa, but no reaction was noted when Black-headed song was played to either of the other two. (Baumann, 2001)

A summary of the typical vocalisations of other oriole species is given in Table 41.

Species	*Scientific name*	*Typical songphrase*	*Typical calls*	*Comment*
Brown Oriole	*Oriolus szalayi*	*titu-witu-whiew tip-top-tertertu.*	*(prrr)-hiiewee trr-ooouееi*	mimics other species
Dusky-brown/Moluccan Oriole	*O. phaeochromus*	not recorded		
Grey-collared Oriole	*O. forsteni*	not recorded		
Black-eared Oriole	*O. bouroensis*	not recorded		
Olive-brown Oriole	*O. melanotis*	liquid whistles, ti-ti-lu-i a long ringing ti-u.		
Olive-backed Oriole	*O. sagittatus*	trembling whistle *orry-orry-ooo*	sneezing *chee-et*	mimics other species
Yellow or Green Oriole	*O. flavocinctus*	a repeated *"shut-you-up" cholonk cholonk*	sneezing call	
Dark-throated Oriole	*O. xanthonotus*	repeated *eeeeee-yopp peu-peu-peu-poh*		
Philippine Oriole	*O. steerii*	repeated *per-jek kok-kok-kok-kok*	*da de hooou hit-di-di hooooi*	
White-lored Oriole	*O. albilorus*	mournful slow *chow-wooooo*	repeated *chup-chup-chup*	
Isabela Oriole	*O. isabellae*	not recorded		
Slender-billed Oriole	*O. tenuirostris*	a series of whistles based on *gri-gorii*	squeaking and chipping calls	
Black-naped Oriole	*O. chinensis*	a series of whistles based on *gri-gorii*	squeaking and chipping calls	
Green-headed Oriole	*O. chlorocephalus*	liquid *wee-ooo*	whining *heee-eeee-aaa*	
Sao Tomé Oriole	*O. crassirostris*	*ouu-huu hui-yiioouh he-ou-hu*	high pitched *keeea*	
Western Black-headed Oriole	*O. brachyrhynchus*	waw-hah and many other variations	wah-chop *tututu-weeah*	
Abyssinian Black-headed Oriole	*O. monacha*	*wocheelywo cheelowah wokachilly*	*graaa-graaa*	
Mountain Oriole	*O. percivali*	clipped *tyee-who slurred ti-tyao-hoh*	*weeka-ku-weeu weekla-wee-er*	male/female duet
Black-winged Oriole	*O. nigripennis*	repeated *hoo-whee slow ho- hee-her-koher*	*teeyup, co-wah*	male/female duet
(Asian) Black-hooded Oriole	*O. xanthornus*	*yu-HU-A-yu tu-HU-ee*		male/female duet
Black Oriole	*O. hosii*	clear whistles with downward inflection		
Black and Crimson Oriole	*O. cruentus*	*hee-hee-heehiew*	mewing call *teee-u kek-kreo*	
Maroon Oriole	*O. traillii*	harsh *kee-ah* followed by *pi-io-io*	a miaow-like *nyaooooow*	
Silver Oriole	*O. mellianus*	not recorded	a cat-like call	

Table 41. Comparison of songs and calls of twenty-four other *Oriolus* species. Created using data supplied by Peter Jones and Bruno Walther (which will appear in their forthcoming text on the genus in *Handbook of Birds of the World*; reproduced with permission of the publishers).

CHAPTER THIRTEEN

Population: past, present and future

HISTORICAL RECORDS

As with all species, the history of the presence of our subject birds must depend on how far one can go back. The earliest reference discovered so far seems to be as discussed in Burton (1995). Burton suggests that by the late 17th century we had lost breeding Hoopoes, Golden Orioles and Bee-eaters. There is a lack of evidence about the status of Golden Orioles in the 18th century, and it is not until the mid-1800s that records start to appear in county avifaunas in any meaningful way. Between 1850 and 1890 records that appear reliable are to be found in county avifaunal publications.

In *Birds of Northamptonshire* Lilford (1883) remarks that they 'can hardly be considered rare between 1860–1865'. The early Kent data all appear to refer to an area in the east of the county between 1836 and 1875. The experience the Golden

Region	*Number*	*Breakdown*
South	20	Sussex 4, Surrey 8, Kent 8, Dorset 1, Hampshire 1
West	13	Devon 2, Glamorgan 5, Gwent 6
East	9	Essex 2, Norfolk 2, Hunts 5
Central	7	Herts 3, Oxfordshire 1, Rutland 1, Northants 1, Worcs 1
North	5	Cumbria 1, Yorkshire 1, Scotland 3

Table 42. Distribution of confirmed and possible breeding 1836–1987.

Oriole Group has in proving breeding even when putting in coordinated effort shows that there well may have been more during this and perhaps any period. Early 20th-century records were again spasmodic, until the start of regular breeding in East Anglia in 1966–67.

Some 54 records have been gleaned from various sources (details are given in Appendix 8). Inevitably, the data are not entirely reliable, but a reasonable picture of the oriole's distribution in the period 1836–1987 can be built up (Table 42).

An additional record for Sussex is of an observer who saw 14 in a thorn bush at Henfield Common and 'got within 40–50 yards'. Unfortunately, the record is undated, making the observation of little value. Of the 54 records, 25 involve birds which bred or attempted to breed; 29 involve birds present in the breeding season, involving sightings of pairs or males singing. The status of the 54 records is given in Table 43.

Collecting of eggs, nests and specimen birds was common in the early part of the period covered, so it is heartening to note that, despite eight nests having failed through human agency, seven were successfully monitored by the finders. A poignant example of human predation involved a nest at Sandwich, Kent in 1851: the report notes 'nest found by gamekeeper and young fed to ferrets'.

	Bred/attempted to breed (25)	*Birds present (29)*
Outcome unknown	9	10
Fledged young seen	7	
Probably bred		16
Adult(s) shot	3	3
Eggs taken	5	
Failed, reason unknown	1	

Table 43. Status of records of Golden Oriole, 1836–1987.

CURRENT SITUATION IN BRITAIN

When discussing population and population trends of orioles in Britain, it is important to put the figures into perspective. According to BirdLife International's publication *Birds in Europe* (Birfield & van Bommel 2006):

> *Oriolus oriolus* is a widespread visitor to much of Europe, which accounts for less than half of its global breeding range. Its European breeding population is very large (>3,400,000 pairs), and was stable between 1970–1990. Although there were declines in a number of countries – notably Turkey and France – during 1990–2000 . . . the vast majority of populations in the east of its European range, including key ones in Rumania and Russia, were stable, and the species declined only slightly overall. Consequently, it is evaluated as SECURE.

A summary of their figures (Table 44) puts our northern populations into perspective. Even allowing for the inevitable errors in population estimates, it is nonetheless clear that the Eurasian Golden Oriole is at the limit of its range in Britain, the Low Countries and Scandinavia. Moreover, the BirdLife International figures are mostly from estimates in the early 1990s, since when we know that further declines have taken place in our region. Most striking are the figures from Denmark (Fahlberg 2004). In the 1980s there were 100+ pairs; this declined to 50 in the 1990s, four in 2004 and just two in 2007.

It is not easy to make categorical statements about the breeding population of orioles in Britain. We know that the figures collated annually by the Rare Breeding Birds Panel are subject to some error, partly because of the reluctance of one or two counties to submit their records, and partly because of the intrinsic difficulty of proving breeding, even in such a rigorously executed study as the one carried out by the Golden Oriole Group since 1987. Table 45 summarises the results for the fenland study area 1987–2005, and shows starkly, especially in columns C and D, the gradual decline in the number of birds occupying sites in the last two decades.

For the years before 1987, and in other parts of Britain, the RBBP annual reports, published in *British Birds*, are the principal source of information. It is clear from the table that breeding outside the fenland basin is sporadic at best.

FLUCTUATIONS IN POPULATION

Oriole populations have a tendency to fluctuate even within countries where they are regarded as common. This was noted in Hungary by Szep & Nagy (2002), and Feige (1986) says 'Local differences in subsequent years may be as much as twice the average frequency of the species.' He analysed data from 16 sources. This

Region	*Breeding pairs (minimum)*	*Breeding pairs (maximum)*
Russia, Poland, ex-SSRs	1,406,000	3,620,000
Central Europe, Balkans	1,521,000	7,768,000
Western Europe	390,000	1,070,000
Baltic States & Finland	34,000	74,000
Holland, UK, Scandinavia	5,670	8,946

Country	*Minimum*	*Maximum*	*Country*	*Minimum*	*Maximum*
Russia	1000000	3000000	Netherlands	4000	5000
Ukraine	140000	190000	Belgium	1600	3800
Belarus	120000	160000	Sweden	50	100
Poland	80000	150000	Denmark	13	19
Azerbaijan	50000	100000	UK	7	22
Moldova	15000	18000	Norway	0	5
Armenia	1000	1500	**TOTAL**	**5670**	**8946**
Ukraine	7	22			
TOTAL	**1406007**	**3619522**	Romania	680000	890000
			Turkey	350000	6000000
France	150000	600000	Hungary	190000	240000
Spain	150000	200000	Bulgaria	100000	300000
Germany	40000	90000	Serbia/Montenegro	100000	140000
Italy	20000	50000	Croatia	50000	100000
Greece	20000	30000	Macedonia	20000	35000
Portugal	10000	100000	Czech Republic	8000	16000
Luxemburg	100	150	Slovakia	7000	15000
Liechtenstein	8	12	Slovenia	4000	8000
TOTAL	**390108**	**1070162**	Austria	6000	12000
			Switzerland	1000	2000
Lithuania	15000	25000	Albania	5000	10000
Latvia	10000	30000	**TOTAL**	**1521000**	**7768000**
Estonia	7000	15000			
Finland	2000	4000	**WORLD TOTAL**	**3356778**	**12540608**
TOTAL	**34000**	**74000**			

Table 44. BirdLife International population estimates (by country) of breeding Golden Orioles in regions of the Western Palaearctic. From Birfield & van Bommel, 2006.

compared numbers breeding within the sourced area with weather, mean temperatures and rain during May/June the previous and following years. Temperature and rain in June of the preceding year have a significant influence on the success of the year that follows. Wet and cold in any one year leads to higher death rates of nestlings than in a dry, warm year, so there are less to breed in two years hence. There are still older pairs to breed the following year and their success may be different if the year is warm. He concludes that a warm and dry May in the

A *Year*	*B* *Sites surveyed*	*C* *Sites with orioles*	*D* *C as %*	*E* *No. of pairs*	*F* *No. fledged*	*G* *Definitely bred (pairs)*	*H* *Probably bred (pairs)*	*I* *Sites with birds present*
1987	39	37	95%	27	20>44	21	6	10
1988	32	21	66%	17	15>20	15	2	4
1989	36	23	64%	18	10>22	13	5	5
1990	30	23	77%	14	11>24	8	5	10
1991	32	21	66%	17	15>20	9	5	7
1992	37	26	70%	20	17>30	13	7	6
1993	39	29	74%	19	31>37	11	8	10
1994	57	30	53%	12	5>25	6	6	18
1995	60	27	45%	17	11>58	10	17	33
1996	69	21	30%	10	16>45	8	2	11
1997	48	20	42%	10	13>44	8	2	10
1998	40	20	50%	12	15>44	11	1	8
1999	51	16	31%	10	13>40	5	5	6
2000	43	10	24%	8	3>7	5	3	2
2001	34	11	31%	4	2>5	3	1	7
2002	37	8	22%	8	4>8	4	0	4
2003	32	10	33%	6	9>20	5	1	4
2004	39	8	21%	5	3>11	3	2	3
2005	40	8	20%	4	4>9	2	2	4

Notes
A Site coverage for 1987 is an estimate
B The total number of sites available for surveying has varied
C Number of sites containing orioles (Categories 1, 2, 3)
D C expressed as a % of sites surveyed in that year
E Combined figure for Cat 1 (definite) and Cat 2 (probable)
F Max number of young fledged using the Milwright Index (see p. 00)
G Category 1, proved breeding
H Category 2, probable breeding
I Category 3, bird(s) present at least once, no further information

Table 45. Breeding success in the fenland study site, 1987–2005.

year of survey brings in an increased number of birds. During spells of bad weather fewer pairs are formed and a lower frequency of nest-building is attained. In periods of longer rainfall even completed nests tend to be deserted. Because it is in the nature of orioles (for males at least) not to breed until the second year after birth, the weather two years previously has the greatest effect on possible breeders and fecundity. The weather conditions in May and June in any given year also have an effect on insect availability the following year. So at a time when there may be more birds there may be less food and *vice versa*. He concludes that 'the influence of weather on older orioles may be mediated by food supply as well as the younger ones, but perhaps a year of non breeding for reasons of a wet May leads to a stronger impulse to succeed in the following year.' There appeared to be a similar phenomenon in the fenland study area. In and after 2001, it was noticeable that

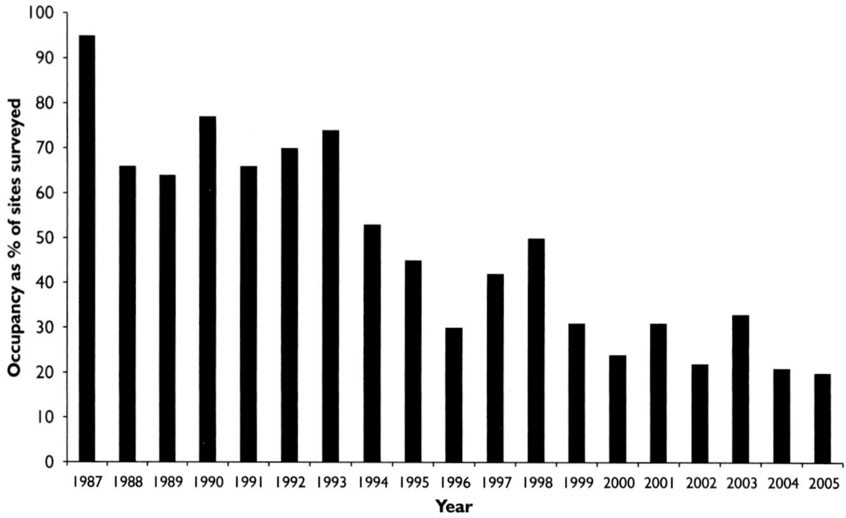

Figure 34. Site occupancy in the Golden Oriole Group Fenland study area, 1987–2005, combining the three recording categories (1 definite breeding, 2 probable breeding, 3 birds recorded but no further evidence).

fledged bird numbers per nest were lower than before. Why other passerine species seem to be less affected by these weather fluctuations is not clear, but it may be that orioles are more vulnerable because their nests are so often placed at great heights in the trees they use, whether in plantation blocks or in more open shelter belts.

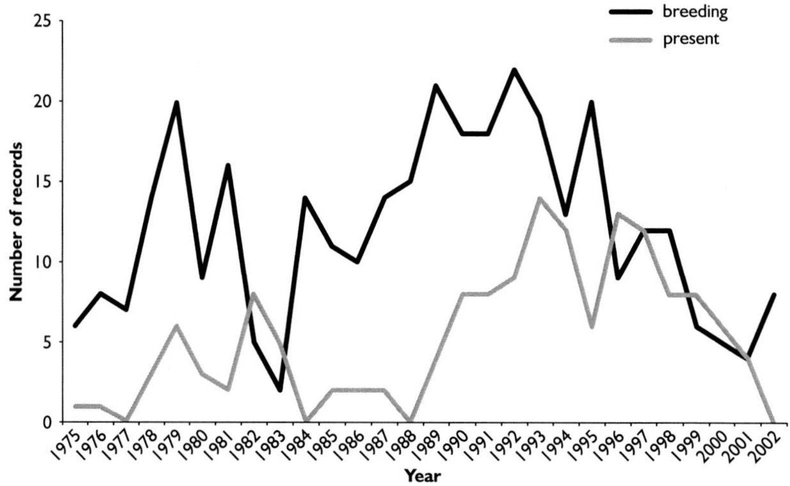

Figure 35. Breeding data for the UK 1975–2002 as published by the Rare Breeding Birds Committee in *British Birds*; 'breeding' includes 'definitely bred', 'probably bred' and 'possibly bred'.

CAUSES OF THE DECLINE OF ORIOLES IN NORTHERN EUROPE

Much of what follows is of necessity speculative. Many species experience cycles of decline and expansion. In some cases the cause is easy to identify, as with the Barn Owl for example, whose three- to four-year cycle exactly mirrors the cycle of their principal prey item, the Short-tailed Field Vole. Sometimes, the cause is catastrophic, as for example the crash of the population of Dartford Warbler in the severe winter of 1962–63, from which it took many years to recover. A similar catastrophic explanation was put forward for the dramatic crash in 1979 of the migrant warbler, the Common Whitethroat, namely the degradation of its winter feeding grounds in the Sahel. Thankfully this species has since achieved a spectacular recovery.

Explanations for most contractions and expansions of range are equally hard to explain in many cases. The Red-backed Shrike and the Wryneck, for instance, used to be regular breeders quite far north in England, where the fact that they had dialect names, 'butcher bird' and 'snake bird' respectively, is a sure indication that they were common birds, and yet their range contracted inexorably from the 1950s and they have now disappeared completely from the British breeding avifauna. Paradoxically, the Red-backed Shrike's range is increasing eastwards in Europe for reasons unknown. Equally speculative are the reasons given to account for range expansion. Recent examples in Britain include Cetti's Warbler, from an isolated record or two in the 1970s to an estimated population of 851 singing males/pairs in 2003 (Rare Breeding Birds Panel, 2007); Little Egret, pushing farther northwards each year from their Hampshire/Dorset base; and Mediterranean Gull, once confined to Needs Oar Point in Hampshire, but now in several localities. Most spectacular – even if the bird itself is not – is the colonisation of western and northern Europe by the Collared Dove, from its original base in Turkey and Southeast Europe, which began in the 1930s. It bred for the first time in Britain in 1955, since when it has become a familiar bird in every park and garden in the land.

The four factors most frequently adduced, singly or in combination, to account for substantial population change are:

1. Changes in agricultural practices
2. Habitat degradation or improvement
3. Climate change
4. Changes in availability of food prey items.

1. Changes in agricultural practices

There can be little doubt that the decline in a number of farmland bird species in Britain, such as Yellowhammer, Tree Sparrow and Corn Bunting, is directly attributable to changes in agricultural practice, including stringent hygiene regulation, the disappearance of winter stubble, and so on. There is little evidence to suggest that orioles are affected by this, although the decline in Belgium has been attributed in part to the decline in large invertebrate prey items as a result of intensive use of pesticides. In the case of Belgium, habitat loss and the uprooting of orchards are more likely to be key factors.

2. Habitat degradation or improvement

The fortunes of the Wood Lark in England are closely linked to the brief periods of suitable habitat in coniferous plantations after old trees are felled and young ones planted. The loss of trees with suitable cavities would have been devastating for Barn Owls but for the provision of alternative accommodation in the form of nestboxes (it is estimated that 80% of Barn Owls in Britain now nest in artificial sites; Peter Wilkinson, pers. comm.). What we can say in the case of orioles in Britain is that loss of habitat cannot be a cause. Put simply, there is a lot more suitable habitat than there are orioles to occupy it. Elsewhere, changes in the fortunes of orioles can be more directly linked to changes in habitat. In The Netherlands, for example, the oriole was considered a common bird in the first half of the 20th century, breeding in tall standards in orchards. When these were uprooted and replaced with half-standards to make harvesting easier and cheaper, the oriole population crashed. Luckily, the concurrent practice of poplar planting in newly created polders provided a good alternative habitat. Even so, orioles have declined in The Netherlands from 7,000–10,000 pairs in the 1980s to half that number in 2005 (Bijslma *et al.* 2001). One factor in this decline may be the unsuitability and consequent removal of some of the cultivars planted (the variety rocket, for example, was very vulnerable to wind-snap). The widespread planting of other species such as oak, ash, elm and maple in place of poplar may have merit, but they take longer to become mature enough for orioles to make use of them. Habitat degradation or change cannot account for all declines, however. In Finland, for example, orioles breed in birch in that country's extensive boggy forests, a habitat that has hardly changed at all, and yet the population has declined dramatically.

A powerful hypothesis put forward by Sabine Baumann (1999) in the course of her work on orioles in northwest Germany is the effect of habitat fragmentation on the viability of breeding populations. There are numerous examples in the literature of the relationship between habitat fragmentation and decline of species. The mechanism appears to be that fragmentation isolates subpopulations and thereby weakens the gene pool. A well-researched example involves the canyons of San Diego in southern California, many of which have been cut in half by highways, so that, despite the pride that the city takes in its canyons, a number of important species of flora and fauna have declined in these remnant 'half-canyons'. Baumann's proposition is that orioles need to breed in concentrated groups in order to provide a nucleus or pool from which the birds can in due course radiate outwards to neighbouring suitable nesting sites. A corollary of this is that, in poor breeding seasons, those young that do fledge are available to plug the gaps left by the death of older birds, thereby maintaining the core recruitment zone, and in good seasons are available to radiate to other sites. In a similar vein, Baumann notes that the breeding season of orioles in Western Europe takes about 10 weeks, and early breeding in coordination with caterpillar cycles is an advantage. Unpublished data from Lower Saxony show earlier breeding and higher breeding success in those subpopulations where several pairs form well-established 'semi-colonies' in habitats that give room for more than three breeding pairs in comparison with more isolated breeders.

3. Climate change

Perhaps the most controversial issue concerns climate change and concomitant global warming. We have seen how vulnerable Britain's orioles are to vagaries of the weather, but it is much more difficult to be categorical about the nature and effects of long-term climatic trends, whether proven or hypothetical. Of course, climate change and global warming tend to be manna to the media, always eager for another dramatic story. Even the broadsheets are not immune. In 2001, *The Independent* claimed that the fact that a record number of orioles had been seen in May on the Isles of Scilly off the south-west coast of England was an indication that Britain was becoming a more attractive place for this quasi-tropical bird. Ironically, in that year fewer bred in the fenland basin than in any year since the year the Golden Oriole Group surveys started. What we can say with some confidence is that the combination of mild, wet winters with warm and wet but unstable summers which we have experienced in northern Europe in recent years has had measurable effects on flora and on invertebrate fauna, with inevitable consequences for the animals which depend on them for food. There seem to be a number of effects:

- Invertebrate prey becomes available earlier (or sometimes later). If summer visitors like the Golden Oriole do not change their arrival time and breeding season accordingly, they will be out of synchrony with their food supply.
- Invertebrate prey does not appear, or are in short supply, as a result of adverse weather conditions affecting their breeding cycle.
- A contraction, or sometimes an expansion, in the breeding range of some invertebrates.
- Synchronisation with prey. A good example of the synchronisation issue comes from a 17-year Dutch study of Pied Flycatchers, where prey, mainly Lepidoptera species, is emerging earlier but the arrival time and breeding season of the birds has not changed. While the connection between these two events is not proven, it seems a plausible explanation for the fact that in some years as many as 90% of nests failed (Both *et al.* 2005).

4. CHANGES IN AVAILABILITY OF FOOD PREY ITEMS

The breeding cycle of invertebrates, particularly of Lepidoptera, is very sensitive to variations in weather. The mechanism is easily described: in winter, when many moths and butterflies are in the larval or pupal stages, they are vulnerable to viral and fungal infections – cold winters kill off these sources of infection; mild, wet winters do not. Thus, the recent run of mild, wet winters, whether a symptom of some long-term climate change or not, have reduced the availability of invertebrate prey. In Britain, much of the data on this comes from systematic moth trapping, both professional (e.g. Rothamstead Experimental Station, with its network of trapping stations around the country; and the Centre for Ecology and Hydrology) and amateur, with an increasing number of moth-trapping enthusiasts every year. It appears that the rate of survival is better in the case of species which either overwinter in the egg stage or hibernate as imagines (Conrad *et al.* 2004). Of the prey species taken by fenland orioles, one, the Herald Moth, overwinters as an adult, and another, the White Satin Moth, overwinters as an egg; both are therefore more likely to be available in quantity. If this is so, it would suggest that availability of prey is not a key factor in the bird's decline. On the other hand, Conrad *et al.* (2006) state that invertebrate species to be found in woodland are declining in deciduous trees while those in conifers are increasing, which of course is bad news for orioles.

Our region is in a warm period, although whether this is just another climatic cycle or whether it is evidence of a permanent and sinister global warming is still uncertain. What is clear is that the ranges of many flora and fauna are expanding or contracting, for whatever reason. For example, in 2006 the UK had its warmest year since records began, and the months of October and November were particularly so. This certainly had an effect on many insects, staying on the wing later and sometimes having extended breeding periods. This may well have an effect on their normal breeding times in future. New insect species that have arrived in Britain in recent years and are spreading northwards include Odonata such as the Small Red-eyed Damselfly, which now breeds here, and the Lesser Emperor, which has been recorded more frequently each year and finally bred at a Cambridgeshire site in 2007. Some authorities link this increase in dragonfly species with the expansion of the Hobby, once a rare breeding bird in Britain. At the time of writing, however, there is no evidence that there has been any increase in the number or range of invertebrate species likely to be of value to orioles.

To summarise, the fact is that orioles are in decline in northern Europe but more or less stable in other parts of their range. The reasons for the decline are mixed and not always clear, but undoubtedly involve one or more of the factors discussed above: changes in food supply, vagaries of climate, and degradation of habitat. The fate of fenland orioles is undoubtedly linked to the fortunes of orioles in the Low Countries: it has long been supposed that our oriole population is not self-sustaining, but that it depends on recruitment from The Netherlands and Belgium and possibly northern Germany. If this is so, then the future, at least in the short term, does not look promising. What makes it so difficult to predict future trends is the fact that, in common with many other faunal species, oriole populations have fluctuated in the past, contracting worryingly and then expanding again, often for reasons that are hard to identify. For example, even in the Czech Republic, a good way south of where the most noticeable declines are occurring, there has been a fall from a peak year in 2004, up to which there had been a slow overall rise. PM and DH did note what seemed to be a late arrival of birds in parts of Bohemia in 2006, as pairs were still displaying as though they had just arrived at the end of May.

As far as global warming is concerned, it is facile to assert that an increase in mean temperature will be beneficial to orioles. *Oriolus oriolus* may be 'ein Tropenwaldvogel in Europa' (a tropical bird in Europe), the subtitle of Ralf Wassmann's book *Der Pirol* (2004), but the consequent effect of climate change has been to upset the quantity, variety and timing of emergence of a range of invertebrate prey species, with dire consequences: adults less able to build up a sufficient protein base to come into breeding condition, brood sizes smaller and fledging rates reduced. Even though the oriole is a relatively long-lived passerine, it is difficult to see how our fenland orioles can sustain their numbers in the changed circumstances both here and in the recruitment zones in the north of Europe unless the ecological balance is restored. There is some merit in the thought that British oriole numbers built up in the early stages of climate change, but as more extreme variations occurred in successive years, breeding orioles were adversely affected.

CHAPTER FOURTEEN

Migration

Much useful information on movement and migration in orioles has been provided by ringing over the years. To 2004, 211 Golden Orioles have been ringed in Britain. Of these, the Golden Oriole Group ringed 83 pulli and one adult. Three of the Golden Oriole Group's pulli and the single ringed adult were subsequently controlled or recovered, a remarkable rate for a passerine (a control is a ringed bird that is trapped or otherwise recorded and then released alive; a recovery is a bird found dead). Otherwise, there have been only three orioles controlled or recovered, all within Britain. The information gathered, in addition to biometrics taken at the time of ringing, provides valuable support to visual observation of movement and migration patterns. Sometimes ringing recoveries can mystify rather than clarify, as in the case of a bird ringed on North Ronaldsay, Shetland, on 23 May 1988 and recovered there 'in unknown circumstance' on 23 March the following year. 23 March is a remarkably early spring arrival date, if that was when it was; it certainly would not have bred in Shetland. Ringing data have proved invaluable in mapping migration routes. They also provide fascinating insights into the distances travelled by migrating birds. To take just a few examples:

Ringed	*No. of birds*	*Controlled/recovered*
Camargue, France	1	Republic of Congo (Zaire)
Spain	2	Republic of Congo (Kasai)
Netherlands	1	Republic of Congo (Kasai)
Czech Republic	1	Malawi
Poland	1	Egypt

The recoveries that have best illustrated the migration behaviour of Golden Orioles are those to be found around the Mediterranean and further north in Europe (Figure 36).

It would appear that Golden Orioles move north in spring across a broad front, from their sub-Saharan wintering grounds, across that desert, to arrive at points along the north African coast from Morocco to Egypt. Much of this data is summarised in Baumann (2000b), as illustrated by the ringing recoveries in Figure 38.

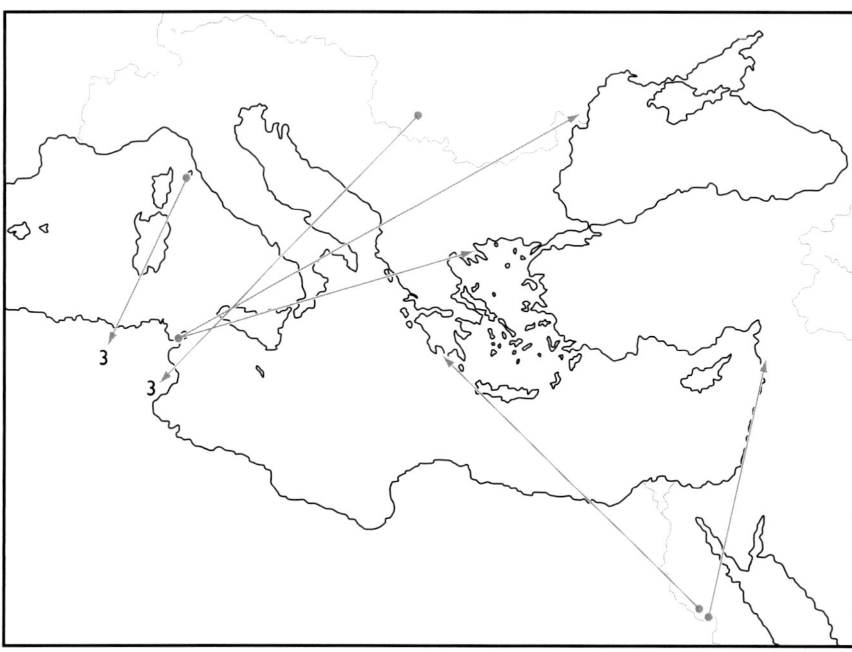

Figure 36. Significant trans-Mediterranean Golden Oriole ringing recoveries. Ringing data presented in the maps in this chapter has been derived from the following sources: BTO (Dungeness), Maria Dimaki (Greece), Pascal Etienne (France), Chris du Feu (Euring), Goodman & Meininger (1989), Heligoland Bird Observatory, Moldovan Istvan (Romania), Societé Jersaise, Sylvia (Poland), Petar Tankov (Bulgaria), Thevenot *et al.* (2004).

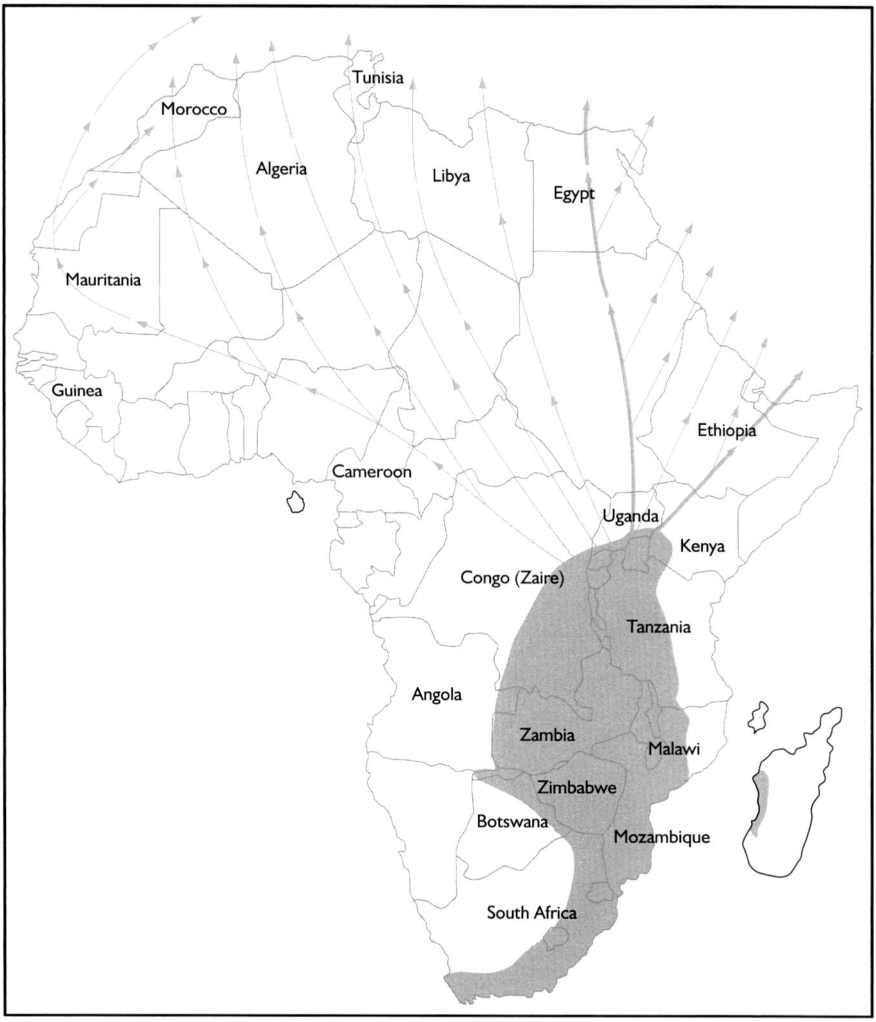

Figure 37. Migration routes taken in spring by Golden Orioles returning to breeding areas.

The birds then proceed north over the sea to Spain, some stopping off in the Balearics (spring only), and in France, Malta and Italy, including Sicily. It is in these countries that observations are most intensive, and it is ringing recoveries like those given below that provide most corroborative evidence. A great deal of spring ringing has been carried out on the island of Capri, and apart from retraps and local controls (some on the nearby Italian mainland) at least three recoveries have been significant. Details of these and others birds controlled or recovered in autumn are given in Table 46.

Figure 38. All European and sub-Saharan African Golden Oriole ringing recoveries.

The routes taken by migrating orioles are further defined by controls or recoveries returning in spring. A number are of orioles shot in spring on Malta which had been ringed as pulli in Hungary, Slovakia or Capri. Other notable records are detailed in Table 47. The Belgian recovery of the Moroccan bird in particular shows how far west the birds can go before turning north-east to their ultimate destinations.

Place of ringing	*Recovered/controlled*
Capri, Italy	Ukraine
Capri, Italy	Greece
Capri, Italy	Egypt
Capri, Italy	Egypt
Corsica	N. Greece
Naples, Italy	S. Greece
Camargue, France	Greece
Cap Bon, Tunisia	Hungary
Cap Bon, Tunisia	N. Greece

Table 46. Birds ringed in spring or summer and recorded or controlled in autumn.

Place of Ringing	*Recovered/controlled*
Romania	Tunisia
Egypt	Greece
Antikythera, Greece	Gavdos, Greece
Morocco	Belgium

Table 47. Notable records of recovered or controlled birds in spring.

RINGING OUTSIDE EUROPE

Birds ringed outside Europe, which are occasionally controlled or recovered, provide further insights into patterns of movement and migration. Examples include:

Ethiopia > Iran
A bird netted whilst on autumn migration in southern Ethiopia, along the Rift Valley, and found dead on breeding grounds near Tehran in Iran the following year.

Egypt > Lebanon
An Egyptian-ringed bird on a southward journey later recovered in Lebanon.

Egypt > Greece
Ringed 23 September 1971, having spent time somewhere in Europe, then migration to Africa again and finally recovered in Greece (Peloponnese) on 6 May 1972.

It appears, therefore, that most western and eastern European breeders migrate north-west on the Nile–Adriatic axis. After rearing their families, they seem to proceed in an easterly, or south-easterly, then southerly direction to Greece, where the overwhelming number of ringing recoveries are when the birds are on autumn, i.e. southerly, migration. This is in complete contrast to the position in Italy and Malta.

It is not certain whether the Ukrainian recoveries were Russian or at least far-eastern European breeders. The fact that a bird ringed in Latvia was recovered in western Crete would suggest that orioles from European Russia take the Greek route. It is also unclear where the dividing line is between the two migratory flows. It could be the Urals; in which case, they could well be the birds observed taking the Middle East route down to south Arabia and Yemen.

MIGRATION ROUTES CONFIRMED BY RINGING

Returning to the question of migration patterns as illustrated by ringing results, the maps and charts accompanying this chapter make it quite clear which direction the European birds take after taking leaving their breeding areas, whether breeders, immature nomads or young of the year. There is a great preponderance for those with an easterly bias, even for those in France: witness the bird ringed on Ile d'Oessant off Cap Gris Nez in Brittany that was caught on Heligoland only 17 days after ringing, a distance of 1,050km. The likelihood is that it turned south from there and joined the main flow of migrating orioles from Germany, Poland, the Czech Republic, Slovakia, Hungary, Serbia and Romania funnelling down to Greece. There it would be joined by others from south-east France, Switzerland and Italy, again all evidenced by recoveries from Greece and the Adriatic coastlines such as the Boot of Italy and Zakynthos.

Antikythira is a Greek island between the Peloponnese to the north and Crete. The ringing station on Antikythira rings more orioles than any of the other Greek ringing stations, but it is worth noting here that, while mainland Greece seems to control or recover more autumn birds, Antikythira gets mostly spring migrants (F. Akriotis, pers.comm), and the figures in Table 48, supplied by a regular ringer on the island, Maria Dimaki, illustrate this well.

It will be obvious that fewer are ringed in autumn than spring, which is strange considering that there have been many recoveries from the Greek mainland in autumn.

	1998	*1999*	*2002*	*2003*	*2004*	*2005*
Spring	52	22	32	31	232	253
Autumn	4	17	nr	1	11	nr

Table 48. Numbers of birds ringed at Antikythira. Date from Maria Dimaki; nr = 5 no ringing undertaken.

As fewer orioles are noted on spring passage on mainland Greece (these being predominantly on the Adriatic islands of Zakynthos, Cephalonia and Levkas), although the breeding population is mainly in the north of the country, the question arises as to where those that land at Antikythira are heading. Maria Dimaki says that they still only trap a small proportion of those that go through the island. The birds have come, no doubt, from the North African coast, but as those birds will have arrived all along that coast, it is not known if they have re-orientated east and are bound for Turkey and the Caucasian countries and on to eastern Russian breeding grounds.

In his treatise *Problems of Mediterranean–Sahara Migration*, Moreau (1961) notes radar-tracking of all spring, mass-migrating passerines over Cyprus reveals that the birds were mostly on a SW–NE course, and that these would have come from the Nile Delta or to the west of it. He also quotes a local birdwatcher's observation of an 'enormous' April night-time movement from Sinai and southern Israel. Conversely, in autumn over Cyprus, again for all species, radar tracks showed general movement to be east of south towards Syria and Israel, and on to the Red Sea. Birds recovered or controlled in Greece on autumn migration, other than Greek-ringed birds, originate from the following countries.

Capri (Italy) has been a regular ringing station used by Swedish ringers over many years, and their results confirm the eastward bias of oriole migration. All their oriole recoveries were at longitudes to the east of Capri, including locations such as Brindisi, Zakynthos (2), Peloponnese (2), Naxos and, furthest of all, western Ukraine (Figure 40).

The station has had spring recoveries from Malta and Sicily, and these have been slightly west of south. There were also spring recoveries from mainland Italy, which

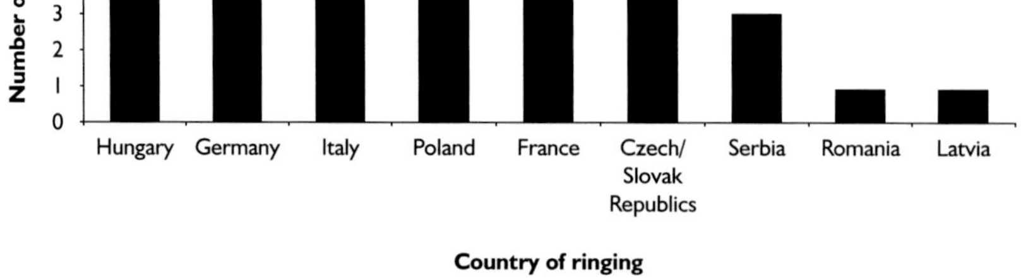

Figure 39. Summary of Greek autumn recoveries of foreign-ringed birds.

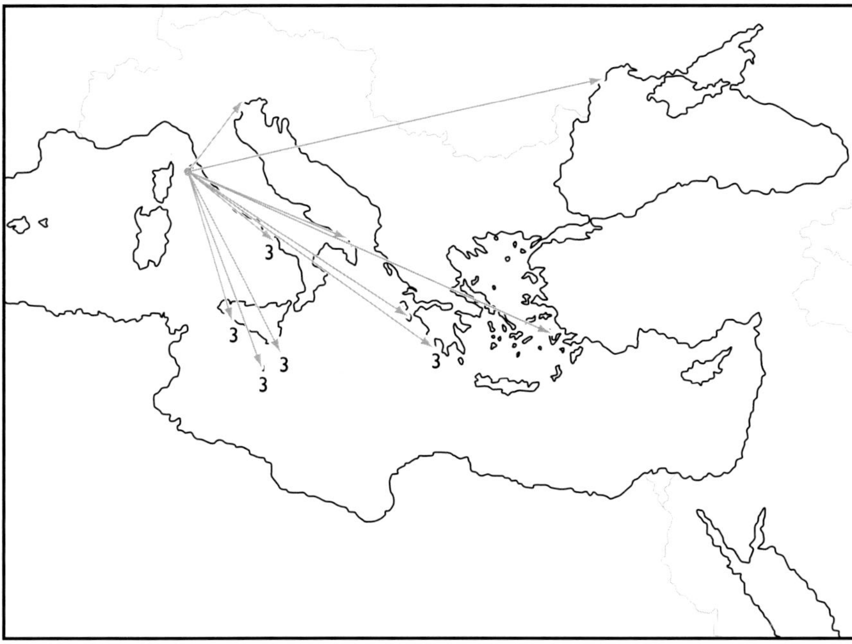

Figure 40. Significant Golden Oriole ringing recoveries from the ringing station on Capri.

were most likely local breeding dispersals. There are positively none from any western direction in autumn migration seasons. Several recoveries along the western Black Sea coast would almost certainly be on a westerly route over Greece. These will have been birds ringed in Capri, Tunisia and Germany and most likely shot in western Ukraine (Euring data), including one found in the Bourgas region of Bulgaria, having been ringed near Warsaw in Poland in 1966 (Donatschev 1976). In August and through September on the 'bottle nose' Bourgas peninsula on the Black Sea, the numbers of orioles stopping off there on autumn migration is described as 'in their millions'. As it is fairly close to Greece in migration terms, it is surprising that similarly large numbers are not recorded there. In fact, despite a large number of autumn recoveries emanating from there, Handrinos & Akriotis (1997) believe that not many are actually observed, so would it be correct to assume that majority of the 'millions' of birds passing through Bulgaria fly over Greece non-stop to the Egyptian/Sinai coast.

That Greece is an important staging area for autumn migrants is confirmed by the number of Golden Orioles ringed elsewhere, mostly in the breeding season, and subsequently controlled or recovered in Greece.

INWARD MIGRATION INTO BRITAIN

There is never a year when Golden Orioles are not seen around the coastal counties of the British Isles, but there is little correlation between the number of these and the numbers actually breeding in any one year, with the possible exception of Kent, where the average number of arrivals correlated well with the peak breeding seasons of 1987 to 1994. Moreover, the places where they arrive can differ greatly from one year to another. There is no doubt that since 1968, when the survey of scarce migrants was begun, the Isles of Scilly have received the lion's share at 323 records, with Kent coming second at 239, and the Shetlands third at 168 (Fraser & Rogers 2002) (Figure 41). Although the Kent incomers may well include some that finally breed, it is clear that the birds recorded on the Scillies and Shetland are drift migrants. Bill Oddie records an exhausted individual on Fair Isle hiding in a bothy in open country.

According to opinion in the Isles of Scilly, they mainly arrive on misty days with an easterly wind, suggesting that they are mainly French overshoots. There have been no ringing recoveries to prove this, but one bird ringed in spring on Alderney (Channel Islands), where overshoots also occur, was recovered in Vienne, central France during the breeding season. The Shetland arrival numbers do not coincide

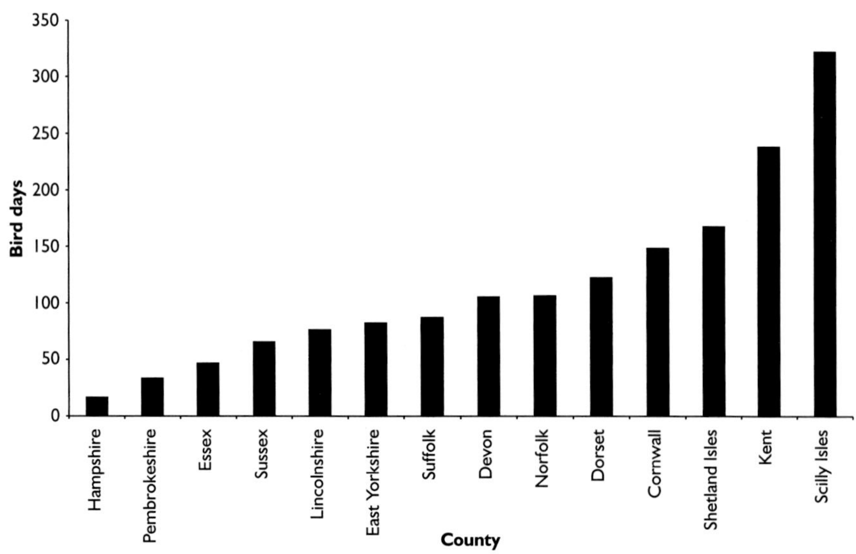

Figure 41. The number of arrivals recorded in various coastal counties in the period 1969-2002 expressed as bird-days; most are recorded from the Scilly Islands, Cornwall and Kent. After Fraser & Rogers, 1969–2002.

with those in more southerly counties, so it is possible that they are drift overshoots from more northerly European breeding grounds. There is some evidence that Kent birds may end up breeding in East Anglia, in that some orioles mist-netted on the north Kent coast have been seen to cross the Thames estuary northwards on release. But of course they may have come from Belgium rather than France.

First recorded arrival dates vary from year to year, presumably dependent on weather conditions determining the rate of progress northwards to breeding grounds. The earliest on record, apart from a doubtful March record, is 1 April (Cornwall). In some years, stragglers are reported as late as 28 July (Devon), although this could well be a wandering non-breeder which in fact arrived earlier.

Earliest and latest arrival dates for four counties are:

County	Earliest	Latest
Cornwall	1 Apr	3 July
Devon	15 Apr	28 July
Scillies	20 Apr	–
Dorset	30 Apr	30 May

Data from the Scilly Bird Reports for the years 1986–2000 give an indication of the modal recorded arrival dates, detailed in Table 49.

The peak arrival time is in late April, with a second peak in the third week of May, almost certainly representing the arrival of females and immature non-breeders.

If, as we believe, our fenland oriole population is maintained by recruitment from the Netherlands and Belgium, it would be logical to expect that arrivals would be regularly recorded along the Suffolk coast. Surprisingly, Suffolk has recorded relatively few. We must assume, therefore, that either they fly direct to their breeding sites or they re-orientate having looped, so to speak, via Kent or the north Norfolk coast, where more are seen than in Suffolk. A recovery of a bird ringed at Gibraltar Point in south Lincolnshire, which is roughly on the same latitude as north Norfolk, and which was subsequently recorded breeding at a Cambridgeshire site, makes the loop theory more plausible.

Earliest recorded dates	*Maximum on one day*	*Number of bird-days*
18 Apr–27 Apr	34	211
28 Apr–7 May	31	195
8 May–17 May	15	69
18 May–27 May	33	176

Table 49. Recorded arrival dates on the Isles of Scilly.

OUTWARD MIGRATION FROM BRITAIN

Outward migration is much more difficult to plot. Orioles are little recorded in autumn in any coastal county; those that do occur are probably a mixture of individual summering birds, breeders and disorientated young of the year. These can be recorded in all months to the end of the year. On the other hand, we have had two results from ringing which we regard as highly significant, in that they both involve birds ringed as pulli on 22 June 1993 in nests 500m apart and which travelled in a south-easterly direction, i.e. ready to cross the North Sea and the Low Countries, from where we believe birds are recruited to sustain the fenland population. One flew into a window at Burstall near Ipswich on 10 August (it is now preserved in Ipswich Museum); the other was controlled at Dunwich on the Suffolk coast on 30 August.

Most strange of all is a record from Leighton Moss (Lancashire) of an adult male seen on several dates – and photographed – at the end of December 2005. It was seen on the ground and harvesting small insects from branches. There is the possibility that it was an escaped cage bird, but the available evidence indicates that orioles are difficult to breed in captivity. Only two captive breeding attempts have been found, both in Germany, so this record remains a mystery.

OVERSHOOT AND DRIFT

Of all the arboreal passerine migrants, Golden Orioles are probably the species most often inclined to overshoot their confirmed breeding range during spring migration. The considerable number of migrants arriving on the Scilly Isles are undoubtedly overshoots from the westerly breeding populations in France and possibly Iberia. Two ringing recoveries may corroborate this supposition: of two birds ringed on Jersey in the Channel Islands, the first was a female recovered on Sark with a brood patch, the second was a male ringed on 21 May 1984 and recovered in its breeding area in Vienne, France, on 18 June 1986 (R. Long, Société Jersaise, *in litt.*). Other Franco-Iberian overshoots have occasionally also been recorded in Cornwall, on Lundy in the Bristol Channel, and in Wales. Even Ireland gets some in most years: up to 1965, there had been 60 records (in one case, five arriving together); since then, 36 more have been recorded, with up to 12 in 1979. The majority of records appear to relate to immature males (judged on plumage and song). It is impossible to say how many females are involved in overshoot and drift, because they are easily confusable with green males and because they are not likely to sing or call.

Arrival of drift migrants and overshoots is highly correlated with certain weather conditions. On the Isles of Scilly, as noted above, they usually arrive during anticyclonic conditions when the light winds are easterly. This is also when other species such as Hoopoe and Bee-eater are likely to turn up in southern coastal counties of England.

Orioles sometimes travel considerable distances over stretches of sea. This is illustrated by the fact that several reach Iceland, over what must have been a hostile crossing. There are even records of orioles on the Faeroe Islands. Between 1965 and 2000, nine orioles found their way to Iceland, most of them making landfall on the south and east coasts of the island, as one might expect, although one did reach the north coast. The numbers found increased after 1973, but it is hard to say whether this increase was because more actually found their way there, because of increased observer effort, or because it coincided with the period when breeding numbers built up in NW Europe. Birds are still turning up occasionally, with two recorded in 2005. Stragglers have also been recorded along Norway's coastline, as far as latitude 60° N. In Sweden, birds have been recorded as far as the Gulf of Bothnia (66° N), having overshot their breeding grounds in the south of the country. Of the most northerly British records of overshoots, Orkney and Shetland, together with the migration watch point on Fair Isle, have the highest numbers of occurences. It is safe to assume that the source of their supply is either mainland Britain or northern Germany and Scandinavia.

Overshoots have also been noted on the Canary Islands (Bannerman & Bannerman 1965). Bannerman & Bannerman also quote a Mr Harcourt, who described breeding on Madeira in 1851, where a nest was purported to have been built on Funchal. In the museum there, there were mounted specimens collected in 1933, 1944 and 1955. Nowadays it is regarded as a 'scarce but regular passage migrant, more common in spring and on the eastern islands'. There have been five records dating back from before 1903. In the Azores, there is a single record of a female, probably shot on São Miguel island (Clarke *et al.* 2006).

ROUTES TAKEN ON SPRING MIGRATION

The fact that the Eurasian Golden Oriole winters as far south as southern Africa and journeys to breeding grounds as far north as Finland and east to Siberia makes it one of the longest-journeying passerine migrants, and arguably the longest-journeying arboreal migrant passerine. Only the Northern Wheatear travels farther, its range extending to eastern Siberia and Alaska.

As Figure 37 shows, our oriole leaves its winter quarters in sub-Saharan Africa and tracks northwards on a broad front, movement taking place from mid-March to early April. Individual birds have been recorded still in their winter quarters on 14 March in Transvaal (Tarboton *et al.* 1987) and 24 March in Namibia

(Immelmann & Sossinka 1973). Northward progress is indicated by records from eastern DR Congo on 12 April (Curry-Lindahl 1981) and from Cameroon, some 1,000 km further north, on 6 April. The birds are then recorded arriving along the North African Mediterranean coast mostly from mid-April on. In Algeria, quite large flocks were observed annually in the oases of El-Golea, Biskra and Laghoat in the northern desert region and Oued Mya and Oued Sarat, which are wadis, i.e. watercourses liable to dry out in summer (Hartert 1910). As many of the great spring bird migration routes pass over the Libyan Desert, there are a great number of visiting species, and many mummified corpses are found.

Interestingly, at El-Golea in Algeria between 14 and 16 May, regarded as a late date this far south, no mature yellow adult males were recorded: all were females or first-year green males. This tends to confirm the general impression that adult males lead the way by several days at least on the journey north. In fact it is claimed that while most spring migration is west of 19° E, autumn migration is east of 17° E (Isenmann & Moali 1987), with, of course, the odd exception. Not all records agree with this, however, as a ringing station at a northern Egyptian site found no marked difference over a six-year period (Goodman & Meininger 1989). The complete opposite was noted in Somalia, where birds were presumably bound for central Asian destinations. Here immatures were ahead of adult males by some 10 days (Archer & Godman 1961). To the west, near Gibraltar, orioles pass in great numbers, mainly between 11 and 21 April (Finlayson 1991). In addition, the Canary Islands and Madeira, which is even further out into the Atlantic Ocean, record orioles on passage (Bannerman & Bannerman 1965)

In Portugal, singing males have been recorded as early as 1 April. According to some authorities, these, and the Gibraltar passage birds, are a discrete population that winters in West Africa. The evidence for this supposition is slight. There are reports of orioles being trapped and ringed during the European winter period in Cameroon, and migration both ways is now known to occur through Mauritania. There have also been a few reports in migration periods, one from Guinea of three males together in gallery forests near Kindia, April 2003, and one in Haut Niger National Park, 28 February 2005 (African Bird Club 2007) The fact that at one time Eurasian and African Golden Orioles were thought to be different races of the same species might have led to this confusion over a discrete population wintering in West Africa and breeding in Portugal.

Further east, there is a peak passage period in early April through Kenya, where they are noted feeding on fruit as they go (Archer & Godman 1961; Jennings 1981; Brooks *et al.* 1987; Lewis & Pomeroy 1989). Dates of passage recorded at Bahig on the Egyptian coast of the Mediterranean over a period of five consecutive years varied from 3 April to 30 May, with the peak from mid-April to mid-May (Horner 1977). Meinertzhagen (1954) noted substantial numbers on passage at Wadi Natrun and El Faiyum, close to the River Nile and 150km south of Cairo, between 12 April and end of May. Thus, it appears that the migration period is quite prolonged, with adult birds presumably preceding younger birds, including first-year non-breeders. He also said that few birds passed through oases in the Western Desert of Egypt. Golden Oriole passage along the Nile Valley is strong, and at Crocodile Island near Luxor observers over a period of about three weeks could see up to 20 birds a day during migration season, mostly in the spring. They are seen along the western parts of the Egyptian north coast, with as many as 20 sightings in one day (A. Sevin, pers. comm.; T. Haraldson, *in litt.*). The spring migration through the region is very evident but is widespread, and again the birds make use of lush gardens on the coast around Alexandria, where large concentrations are seen. Goodman and Meininger (1989) quote the record, in April 1982, of thousands being seen at El Hauwariya to the south-west of Alexandria. Mary Megalli, a prominent Egyptian ornithologist, makes this comment on finding birds in 2005:

> A very usual record, for time and place: oriole at the new golf course in Sharm el Sheikh, April 23. Though I don't know if the east coast of Sinai figures in records or not; having found nice gardens in Sharm el Sheik, one wonders if such birds will continue up the coast or go inland into the mountains and north. The coastal resorts (all of them – north coast, both Gulf of Suez coasts, and Sinai Gulf of Aqaba coast) feature splendid gardens. The passerines go resort hopping up the coasts, some spend the winter. In these few weeks, especially ahead, not much before, there will be orioles in all the gardens, especially along the Egyptian Gulf of Suez coast. As a friend at Wadi Dom, a condo village with great patches of garden, said one gets 'tired' of orioles in late spring. I logged 84 birds there the year before last [2004].

She has also recorded birds sheltering all day from the intense heat of the sun in the shadow of a small rock in the western Egyptian desert. They spent the time almost immobile on the ground, tucked into the bottom of the rock.

Many more fly over the Sinai peninsula, having departed from the African continent via Somalia (rarely in spring, but many in autumn; Archer & Godman 1961), Eritrea and Ethiopia, and on over the Red Sea, up the Gulf of Aqaba to Israel and Lebanon. There is also passage over the Arabian peninsula through Iraq, where sightings were recorded at a remote pumping station on the Iraq–Syria oil pipeline (Hosking 1970). Others go to Iran and along and across the top of the Caspian Sea to breeding grounds in the more eastern parts of the Russian Federation, as far as the Altai Mountains and western China (Xingjiang), north-west Mongolia and in Siberia as far as Lake Baikal. This is a journey of well over 12,500km in some cases. The birds then continue on a broad front into Europe, as can be seen in Figure 42, supported by ringing data shown in Figure 43.

From the beginning of May, many birds arrive in Malta, but considerably fewer leave after shooters take their toll. From then on, orioles arrive on their breeding territories throughout Italy, France, Belgium, Holland and England, and others through Germany to Denmark and to the few nesting sites in Sweden. Those making landfall in Greece (probably from Italy), where some stay to breed, will

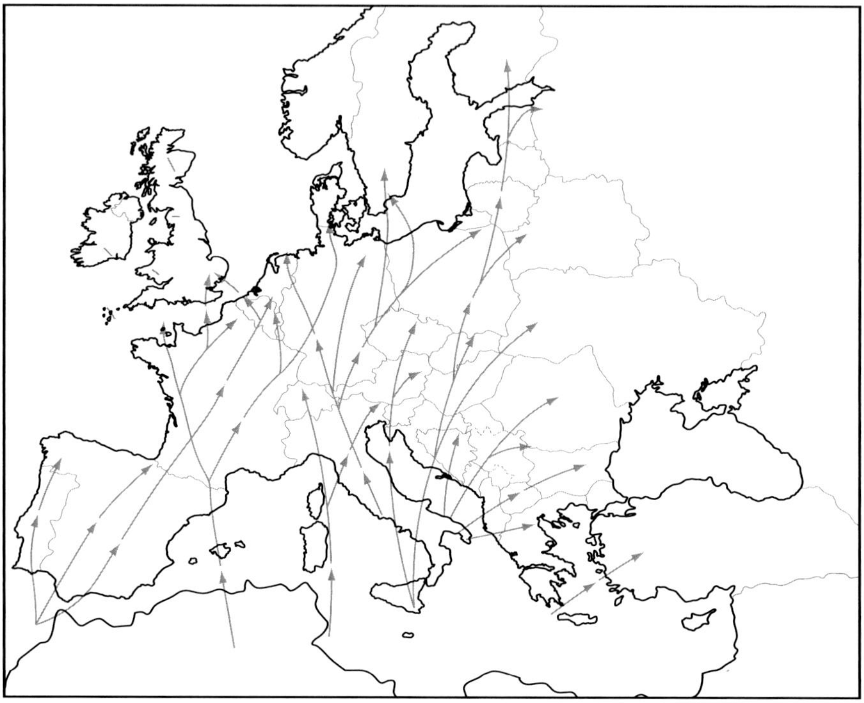

Figure 42. Spring migration routes taken by Golden Orioles through Europe.

move on to the Balkan States, Austria and the central and Eastern European countries: Romania, Bulgaria, Hungary, the Czech Republic, Slovakia and Poland. Further east still, orioles travel from Egypt, sometimes stopping off on southern Greek islands like Crete, Antikythira and Gavdos, where many are ringed, to Cyprus and beyond to Turkey, Moldova, Ukraine, Belarus, Russia and to the Baltic States and Finland, the orioles' most northerly breeding grounds.

Spring migration of Indian Golden Oriole

The Indian Golden Oriole journeys from its wintering grounds in southern India and Sri Lanka north to the Kashmir regions of both India and Pakistan then through the mountain valleys of the Himalayas. Observations of the movements of the birds through the region include one by H.G. Alexander (1974), who, on 1 June, saw a female arrive from the south coming up the valley at Bakrota (2,438m), when she settled on a walnut tree for a time, on the crest of a ridge, and then was seen to continue north. The birds also go via Nepal to southern Xingjiang, and via the Hindu Kush on to Afghanistan, Turkestan and Tajikistan, Kyrgyzstan, central and

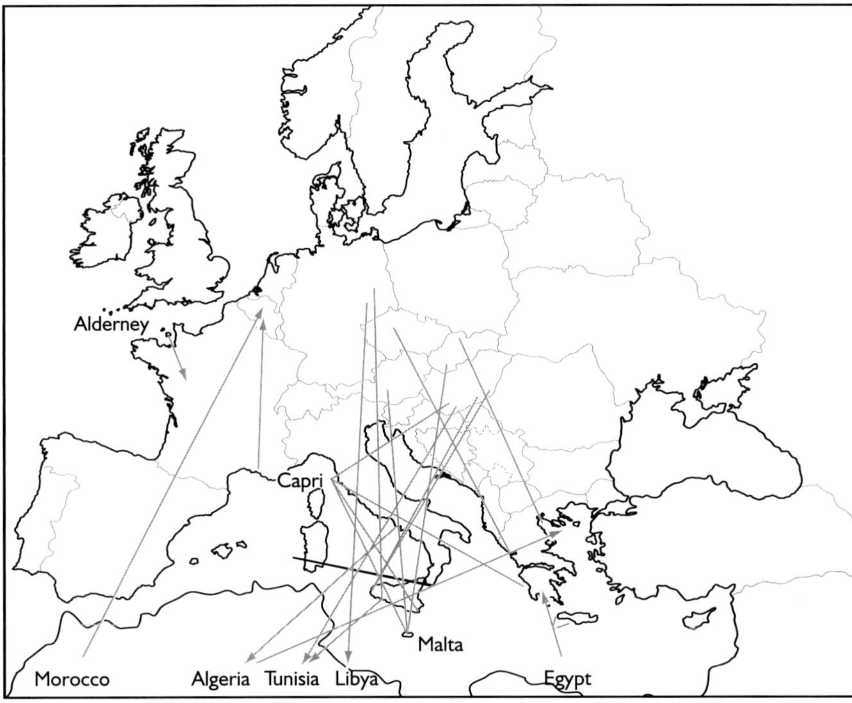

Figure 43. Returning spring migrants ringed in Europe and North Africa after wintering in South Africa.

Location	*Date*	*Reference*
Sri Lanka	March	Ali & Ripley 2001
Islamabad	7 April (three-year mean)	Mallalieu 1988
Khatmandu	Mid-April onwards	Inskipp &Inskipp 1985
West Tien Shan mountain region	27 April (11-year mean)	Gavrilov & Gitsov 1985

Table 50. Records of Indian Golden Oriole on spring passage.

eastern Uzbekistan and to Kazhakstan as far as around Almaty and the far south-east of the country along the River Syr-Darya valley leading to the Aral Sea. In these parts they are liable to meet their congener, the Eurasian Golden Oriole. A selection of dates these birds have been recorded on passage are shown in Table 50.

The record for the western Tien Shan mountain region on the north-west border of Xingjiang province of China and at the Chokpak pass, where the borders of Kazakhstan, Kirgistan and Uzbekistan all intersect, seems to be as far as *O. kundoo* travels; specimens examined further north are of *O. oriolus* which have come from

Africa. However, one of the few ringing recoveries of *O. kundoo* is of a bird ringed in Gujarat in eastern India and recovered in Tajikistan some 2,100km north (Ambedkar 1987). Many do not, in fact, travel so far, but only to the north of India. Some authorities believe it to be resident there, although whether these have wintered even further south within that area has yet to be confirmed.

ARRIVAL DATES

Times of arrival on breeding grounds differ somewhat from south to north. Movement through North Africa seems to range from 19 April to 23 May (Brosset 1956), but some of these are likely to be local breeders, whereas the dates given for Egypt – 23 April to mid-May – refer solely to birds on passage (Horner 1977). The progression of arrival at breeding grounds in Europe is indicated in Table 51.

British orioles only breed regularly in East Anglia but migrants turn up all along the south coast, far more than would account for the number of pairs breeding. Figures for the 'hot spots' show that numbers have decreased since a peak between 1992 and 1994, when totals were 184 and 235 respectively (Fraser & Rogers 2002). This coincides with a drop in numbers of breeding pairs in Denmark, northern Germany, The Netherlands, Belgium and northern France. By 2003 numbers had dropped to 77. To illustrate the decline further, the 10-year average for the period 1990–99 was 132; the average of the first three years of the 21st century, 2000–02, was 90.

During the same periods, numbers peaked in Kent, Cornwall, Devon, and even

Mean arrival date	*Range*	*Location*	*Reference*
22 April	–	Yverdan, France	Salvan 1966a
24 April	–	C. & S. Alps, France	Lebreton 1977
2 May	–	Picardy, France	Etienne & Montel 1986
10 May	–	Thuringen, Germany	Von Knorre *et al.* 1986
14 May	1 May–2 Jun	Eastern England	Milwright 1997
–	1–10 May	Vol'sk, C. Russia	Ushakov 1927
13 May	11–15 May	Kazakhstan	Popov 1978; Gavrilov 1974
–	3–22 May	Tatarstan	Ushakov 1927
16 May	–	Penza region, C. Russia	Ushakov 1927
18 May	15–21 May	W. Siberia, Russia	Johansen 1952
22 May	13–31 May	St Petersburg, Russia	Mal'chevski & Pukinski 1983
end May	–	Perm region, Russia	Ushkov, 1927
26 May	21 May–1 Jun	Finland	Tiainen 1979

Table 51. Arrival dates at breeding grounds.

Date	*Shetland*	*Cornwall*	*Scilly Is*	*Devon*	*Dorset*	*Kent*
1–7 April	0	1	0	0	0	0
8–14 April	0	1	1	0	0	0
15–21 April	0	6	11	2	0	0
22–28 April	0	19	37	5	12	2
29 April–5 May	1	20	63	15	12	64
6–12 May	4	19	72	13	11	39
13–19 May	12	47	129	18	24	70
20–26 May	28	36	104	13	33	45
27 May–2 June	47	30	55	20	18	34
3–9 June	35	6	22	9	8	27
10–16 June	18	2	7	7	10	18
17–23 June	7	1	4	1	2	1

Table 52. Date of peak arrival in coastal parts of Britain.

on the Northern Isles of Shetland and Orkneys. It is also interesting to note here the peak arrival dates, which are roughly the same wherever they make landfall. The years covered in Table 52 are from 1968 to 2001.

Arrival dates at the fenland study area have gradually become earlier. When the Golden Oriole Group first started its collective work in 1987, the mean arrival date was 10 May. However, in 2005, for example, one male bird was on breeding territory on 29 April. The latest date to hand is for 2006, when two birds, again males, arrived on 2 May, suggesting that the trend continues. How this relates to the decline in numbers of breeding birds is not clear. Some of the very early arrivals may be the result of displacement caused by such factors as inclement weather. The latest ones, certainly those after 3 June, are almost certainly immature birds, typically first-year males in green plumage. Females may be included too but are likely to be overlooked.

ROUTES TAKEN ON AUTUMN MIGRATION

Patterns of departure from northern breeding grounds are shown in Figure 44, with supporting evidence from ringing returns shown in Figure 45. Regrettably, in addition to birds controlled by other ringers on migration routes to the south, many recoveries result from illegal shooting and trapping in southern France and around the Mediterranean. Recoveries and controls of birds ringed in northern and central Europe show that they travel in large measure in a southerly direction, through northern Italy, Greece and Turkey, whereas birds in Italy and the Balkans take a an easterly then southerly route before crossing the Mediterranean. Birds that have been ringed in France, both west (Charente-Maritime) and north (Pas de

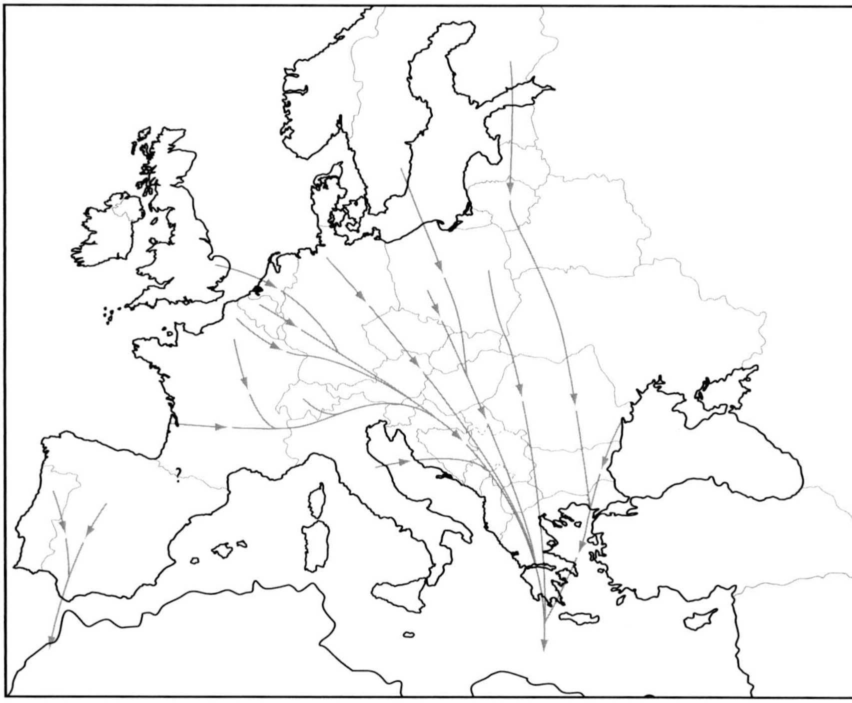

Figure 44. Autumn migration routes taken by Golden Orioles through Europe.

Calais/Picardy/Haut Rhin regions), appear to move to north Italy. Whether they go round the Alps or over them is not yet known. Some type of loop migration seems to occur in the Ligurian Alpine region, relatively close to the northern Mediterranean sea, where migration watches in the autumn record many orioles, in contrast to very few seen in spring.

Those originating from eastern and Siberian Russia and all points north travel via eastern Turkey, having skirted the Black Sea (evidenced by the fact that few are seen in the Crimea, whereas many are observed in north Caucasus/Stavropol region of south Russia; J. Gordon, *in litt.*), including the Kiziltashskaya kosa, a narrow spit of land on the Taman peninsula on the Black Sea coast, which is recognized as an important stopover locality for many migratory passerines. The birds then continue south through Syria, Lebanon, Israel and past southern Arabia, but bypassing northern Iraq. Jordan and Saudi Arabia see birds from the Oman coast to the Yemen, but most birds are assumed to fly at night over the arid desert regions, where there would be little reason for them to make landfall. Then they return south-east over the Red Sea again. After crossing the Bab-al-Mandab straits from Yemen, large flocks have been recorded in Somalia making their way to the Rift Valley, where they join others from Europe that have travelled down the Nile river

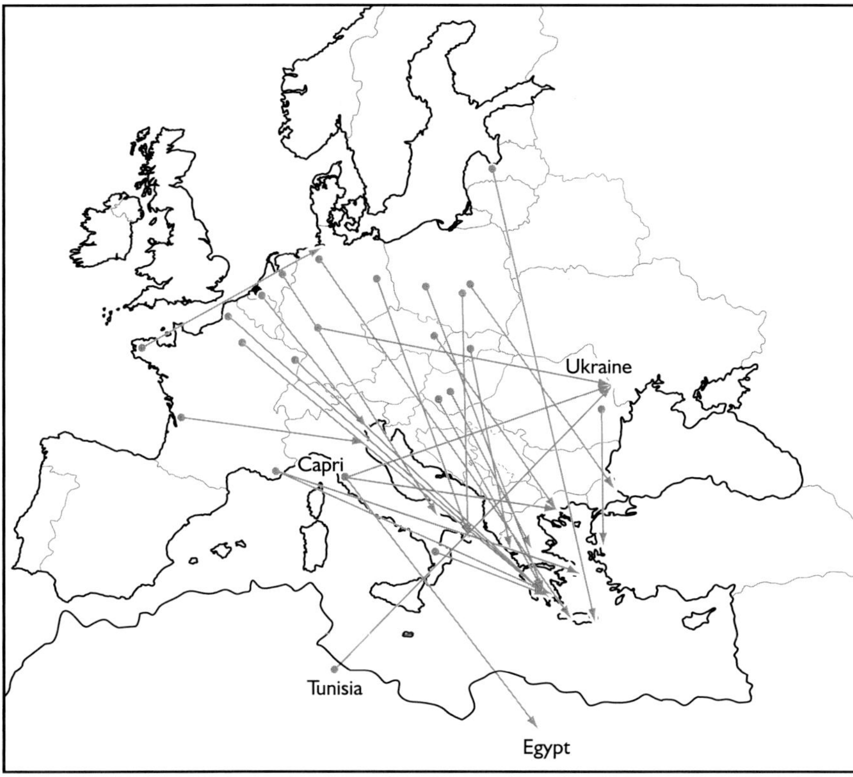

Figure 45. European ringing recoveries in autumn of migrants ringed as possible breeding adults or as pulli.

complex through Sudan (Archer & Godman 1961). Birds have occasionally been downed while migrating over the Indian Ocean, with at least one on Socotra in August 2004 (Richard Porter, pers. comm.). This island is Yemeni territory but lies off the Somalian Horn of Africa. Other birds have been found on Aldabra and other islands in the Seychelles Island group (Frith 1974). All have been in the autumn migration period. There are no reports from the nearby Comoros Island group, but it would be surprising if there were not some because of their proximity to the Seychelles.

The two principal easterly streams of southward migration are shown in Figure 46, which also illustrates the westerly stream of birds from Iberian, Moroccan and Algerian breeding areas, which return south via Tunisia over or round the western Sahara to Lake Chad and to Air in northern Nigeria, before sweeping south-west to join the easterly orioles in their winter quarters. It is important to realise that the routes taken are not randomly chosen. In this instance, for example, the birds are skirting the northern fringes of the equatorial rainforest.

Orioles do, of course, breed in Morocco and Algeria and this may have an effect

Figure 46. Three concentrated migration routes taken by Golden Orioles en route to their African wintering quarters.

on migrating patterns from that area. Whilst there have been no ringing recoveries from Mauritania, to the south-east of those countries there has been quite a lot of ringing of all species, carried out there in spring and autumn in the Sahara project by the Swiss Ornithological Institute. Some 71 orioles have been ringed, amongst other species, and on both spring and autumn migration: 15 birds on spring migration and 56 in autumn in the years 2001 and 2003–05 (Herremans 2003; V. Salewski, *in litt.*). However, this did not reflect the numbers actually observed. During autumn there was a team at Tichit, where in August the first landfall of migrants was a group of about 100 orioles, which stopped over for several days in a small plot of trees. There were definitely not hundreds of oriole in single trees but about 100 orioles at a time in a small forest planted as a windbreaker to protect the oasis from approaching dunes. Almost the only tree species of the plantation was *Prosopis juliflora*, an introduced tree from America, which had no fruit at this time. It was assumed that the orioles were attracted by the possibility of a rest in the shade rather than the prospect of a meal. There definitely appears to be a sizeable population of orioles moving round the western Sahara to wintering grounds in Africa, but if these grounds are in the west, as had previously been believed, then it has not been firmly established. In fact there is much evidence to say that they do not.

Within these descriptions of migration movements in both directions, one can detect the loop phenomenon. Visual observations backed up by ringing data provide the evidence for this. Cramp & Simmons (1993) state that all autumn ringing recoveries are east of 17° E, whereas all spring data received indicate recoveries west of 19° E , leaving a band of only 2° where they overlap. In Africa this line would run approximately 100km west of Beghazi in Libya, up to the Albanian coastline to Budapest in Hungary. There are some striking instances of this looping, such as the case of a bird ringed in Tunisia in May 1955 recovered in September 1957 in eastern Greece, and another ringed in eastern Morocco in April (year not known) and recovered in Belgium in August the year after. Moreau (1961) also collated records from many observers noting migration in both seasons all along the North African coast from Morocco to Egypt. He charted the numbers for each season in eight sections and an interpolation from his data appears in Table 32.

In respect of Tunisia, Moreau remarks: 'The consensus of all the data is that more birds are seen in spring than in autumn, indeed a number of species that are common in spring (e.g. Orioles and Common Whitethroats) are unknown in autumn.' He also comments that this is probably because the coast is especially attractive in spring. In his chapter on the Tripoli section, he in turn quotes Johnson (*in litt.*), who says that in autumn there is nothing for insectivorous birds to stop for. It also seems that there is little fruit here either. For the Cyrenaica section of Libya the information is sparse, probably because of the difficult terrain. Allowing for the fact that the data are mostly based on casual observations, the table provides evidence of arrivals on a broad front along the whole of the North African littoral in spring, but a bias towards the eastern and western ends in autumn, with a greater proponderance at the eastern end.

Yemen is quintessentially a country where many species of birds gather ready to migrate over the Red Sea to Somalia, Djibouti and beyond in Africa. Besides being a bottleneck for large species like raptors, Bab-el-Mandeb (where the crossing is at its narrowest) also sees very large numbers of passerines, including Golden Orioles, where they feed up in vegetated wadis in spring. This area, including Ibb- Dhi-

	SW Iberia, W. Moroccan Desert	*E. Morocco*	*Algeria*	*Tunisia*	*Libya (Tripoli)*	*Libya (Cyrenaica)*	*Eastern Sahara*	*Egypt*
Spring	3	3	3	3	3	3	3	3
Autumn	2	3	0	2	2	3	3	3

Table 53. Records along the North African coast. A rating of 0–4 is given, which relates to roughly the occurrence abundance. 0 rates none seen, 1 for rare, 2 for rather uncommon, 3 is for common, 4 is for very common. From Moreau, 1961.

Sufal, inland at an elevation of 1,500–3,100m, is also very important as a refuelling zone for migrant passerines in autumn, including large numbers of such species as Cuckoo, Bee-eater, Yellow Wagtail, Common Nightingale, Common Whitethroat and, of course, Golden Oriole. Many more fly over the Sinai peninsula, having departed from the African continent via Somalia, rarely in spring but frequently in autumn (Archer & Godman 1961).

Golden Oriole movements in Saudi Arabia, Israel, Jordan and Syria, together with the United Arab Emirates and the States of Oman, Bahrain and Kuwait, also show a degree of loop migration in so far as more birds are seen in spring in the north and east of the region. This also includes regular appearances on islands like Das in the Gulf. In autumn the bulk are recorded from south-west Yemen, where they are preparing to cross the island-strewn southern Red Sea, from the Dahlac archipelago to Bab-el-Mandeb Straits. It is certain that many migrating birds pass through Iraq; there are old records of sightings near oil pipeline pumping stations and others where there were groups of trees. These were mainly spring migration records. Israel has both a spring and autumn passage, but the spring birds seem to take a more western route, stopping off to feed in gardens on figs and dates, than those in autumn. Shirihai (1996) records that, in spring, males are noticeable, arriving 10 days earlier than the following females or immatures, whereas in autumn the reverse occurs, again by a 10-day period. Jordan also records passage in spring but a much stronger one in autumn. The spring passage passes through the desert oases scattered throughout this dry country but many more migrate south in autumn. The birds stop off in eucalyptus trees in late August through to early October, at places like Shaumari in the eastern desert and along the Rift Margin highlands and Northern Highlands. Perhaps the fact that there is little seasonal difference in numbers north or south through Egypt shows this may be the dividing line of migration direction. Birds travelling north mostly following the Nile in spring are bound for Eastern Europe. Conversely those autumn travellers going south are central and Western European breeders (with the exception of the Iberian and Maghreb populations).

The autumn migration routes of central Asian breeders are more difficult to plot. It is not certain what routes south are taken by orioles breeding in Asia. There are only two indicative ringing records: one of a bird ringed in the Rift Valley of Ethiopia as it migrated south and recovered in Iran the following July, and a recovery of a bird ringed in Egypt and recovered in the Lebanon, for which no dates can be yet found. We know that there is a mass southward movement to the Stavropol region in southern Russia, where large numbers are noted feeding on fruit, especially mulberry, before moving on along the foothills of the Caucasus mountain range. There are several records of movement south through western Jordan. Records are also available from further east of substantial passage of orioles through Azraq, which is in eastern Jordan near the Saudi border (Cameron & Cornwallis 1966). This location suggests that orioles may fly down over the Chert Desert and Arabian Desert to the southern Red Sea for their next stop. It has to be admitted that some of this is speculation, but according to various sources, many orioles do cross over the southern Red Sea: they are recorded as a passage migrant in Sanaa, Yemen, in spring but more frequently in autumn (Brooks *et al.* 1987). From here they seem to fly over the Arabian Desert down to Aden and the Dahlac archipelago, where great numbers have been recorded crossing the water and entering Somalia, and many have been ringed along the southern Sudan Red Sea coast in autumn. Archer & Godman (1961) give a very descriptive account of 'a wonderful seasonal passage of the Oriole through this part of Africa' for a short period on the Somali coast. They go on to say when staying at Erkowit in the Red Sea hills that they saw:

> . . . numbers of these immature birds clad in their sombre olive-green and brown garb, which were pouring through daily, had to make the best of the trees and shrubs which grew on rocky barren ground in poor contrast to the shelter they would enjoy at a later stage in their journey on reaching the well-watered, heavily timbered and verdant Hargeisa valley of western Somaliland.

From Erkowit the birds head for the northern opening of the Great Rift Valley. After they have passed through Erkowit there is a period of about 10 days before full-adult males arrive again in large numbers overnight. Again we quote from Archer and Godman's highly evocative text: 'they create vivid streaks of colour as they flash across open ground between heavily foliaged trees, in which it is their habit to conceal themselves.' It seems they will then sing. Birds have been seen on the open savannas near Tug Wujala, where there are rivers and where the birds were seen plunging into the water to catch insects. The period of passage in these parts is 7–27 September, after which no Golden Orioles are to be seen.

By this time, other orioles have also crossed the Mediterranean to Egypt, where they feed on fruit, especially dates. In fact, they have reached pest proportions: flocks of around a 1,000 have been seen at times, as at Bahariya (Goodman & Meininger 1989). Annie Sevin (*in litt.*) noted on 6 Sept 2006 that orioles had been gathering around the Bahariya Oasis area for about a week, and in doing so she made the ominous remark 'where the hunters are waiting'. Bahariya Oasis is set in a depression covering over 2,000km^2 and wildlife is plentiful, especially birds such as wheatears. A small percentage of the total area includes dates, olives, apricots, rice and corn, and during migration the whole area known as the Western Mediterranean Coastal Desert receives vast numbers of Palearctic migrants. The region provides the first available resting and feeding area for huge numbers of individuals of many species after their Mediterranean crossing. Most famous of these are Common Quail and Golden Oriole. As they pass down the Nile they arrive at Khartoum and stop in considerable numbers at the beginning of September to feed on figs and inhabit the lime trees and, as along the Red Sea at this time, are mainly immatures. From here they pass on down the Nile to Uganda. In the Darfur region to the west of Sudan and in Chad, Admiral Lynes is quoted as saying autumn migration is in early September in fluctuating amounts. Males, sometimes singing, had passed through by the middle of October (Archer & Godman 1961).

Farther east still, breeding Eurasian Golden Orioles have been recorded migrating south through Pakistan, for example in Malir. Pakistan's western and southern provinces have seen them on a few occasions, mainly in the months of August, September and October in Karachi and at Chabar in the eastern province of Baluchistan. There has been one spring record, that being at Tharparkar on 16 April. These records cover the years from 1979 to 1983 (Roberts 1991). While most records in India are of the Indian Golden Oriole *O. kundoo*, noted aberrant southward migration of Eurasian Golden Orioles has been observed in north-west Afghanistan, Ladakh, coastal eastern Pakistan and even Sri Lanka (Rasmussen & Anderton 2005). These birds presumably got mixed up with Indian Golden Oriole movements at the same time.

Further evidence for oriole loop migration is provided by the fact that there are places where orioles are common on passage in spring but virtually unknown in autumn. For example, in Malta, where much shooting of birds takes place, large numbers arrive in spring, whereas very small numbers are seen, let alone shot, in autumn migration periods (Roberts 1954; DeLucca 1969). Likewise there are similar records of strong spring but poor autumn passages from Gibraltar, and from Corsica and Sicily, with similar observations from the Balearic Islands.

Even the Canary Islands, some 150–200km out into the Atlantic ocean, see examples, some years in quite large numbers, of birds going north in spring but never of birds going south in autumn. Whitaker (1905) adduced a theory to account for their absence on autumn passage in contrast to their abundance in spring, namely that autumn migration must take place at night, with birds passing over without resting, whereas in spring they arrive from the south in daylight hours and are therefore more obvious. There is no evidence to support this theory.

Stresemann (1948) believes that the Golden Oriole evolved this pattern of loop migration as a result of its evolutionary spread. He says: 'If this is correct then the autumn migration is approximately in the same direction in which the expansion of the species has developed whereas the spring migration shortens this circuitous route to the western breeding areas of the species.'

LOCAL LOOP MIGRATION

The pattern of migration is that the birds move north from Africa into Europe on a broad front (Figure 37), with an easterly band moving to the Asian breeding grounds, whereas in autumn there is a well-defined front leading to eastern and southern Africa, including birds of the more westerly population, which turn south-east after sweeping down the west of Africa. In addition to this macro-loop migration, there are examples of local loop migration, exemplified by, among others, the situation in Kazakhstan (E. Gavrilov, *in litt.*). The Chokpak Ringing Station in southern Kazakhstan has ringed over 500 orioles. Many fewer are ringed at Chokpak in spring than in the autumn, and Gavrilov believes this is due to loop migration following the foothills of the Kirghiz mountain range as they feed up on supplies of fruit as they migrate west to south-west. After leaving this area, they have to negotiate the Kuzyl Kum desert and then, in accordance with what one might call the 'fruit-trail theory' they skirt the northern shores of the Caspian Sea and arrive in Russian territory near the Volgograd region before going south to the Caucasus. Interestingly, Chokpak also rings *kundoo* orioles and is the only station to so do. In contrast to the Eurasian Golden Oriole, many more Indian Golden Orioles are ringed in autumn than in spring. The reason for this is again thought to be because they are following their own fruit food source in a southerly direction along the Altai, Pamir and Tien Shan Mountain foothills, but when they arrive in spring from Indian wintering quarters they are migrating on a broad front. Gavrilov further comments:

> Regarding the difference in numbers of *kundoo* and *oriolus* at Chokpak, the fact is that in spring, deserts are green, with many flowers and insects. Therefore many birds migrate to Siberia and northern Kazakhstan in a direct way and only if they get diverted for any reason (such as adverse weather conditions) do they fly east and turn up through Chokpak pass. But in autumn many birds migrate through forested mountains, which in spring were covered by snow, and avoid migrating through deserts. It is small loop, caused by difference of ecological situations in spring and autumn.

In autumn the birds would move south over land and probably mostly by night down to the north Caucasus (Stavropol) region of Russia. Here there are reports of many migrating orioles feeding up on the numerous mulberry, bird cherry and other fruits, before tackling the mountains proper.

HAZARDS OF MIGRATION

The principal hazards of migration for migratory species include weather, predators, trapping and hunting.

Weather

Migration for any species is a risky business. For instance, birds can be downed by strong winds into inhospitable terrain such as sandy deserts or small rocky islands in wide oceans, only to be picked off while in a weak state by the local predators. Whichever route orioles take, they have to cross one or more deserts. The European breeders have the Sahara to negotiate, while those bound for mid-Siberia have two – the Arabian Desert, and, further on, Kyzul Kum on the eastern side of the Caspian Sea, where sand storms cause many casualties.

Predators

When it comes to crossing open sea, two problems arise. Over the western Mediterranean Sea, Eleonora's Falcon has timed its breeding season to coincide with autumn migration. Inspections of eyries has found that a proportion, albeit a small one, of the total prey consists of orioles (Hedenström 2003), whereas in the eastern section, the Aegean Sea, the tally was much higher from August to October, being the third most preyed-upon species, the first two being Spanish Sparrow and Common Nightingale, out of a total of 14 recorded (Ristow *et al.* 1984).

When the orioles return south-east over the Dahlac archipelago in the Red Sea, another raptor, the Sooty Falcon, takes its toll. Juvenile or immature orioles are the usual victims, for it is their feathers that are found most in nests (Clapham 1964). Other prey species include Hoopoe, and European and Madagascar Bee-eaters. It is probably no coincidence that these are all brightly coloured, highly visible birds. In Britain, the only record of possible predation is of an oriole ringed on Bardsey Island and found dead some days later as part of the prey items in a nearby Peregrine eyrie, although whether it was killed by the Peregrine or collected as carrion is not known (there is a reference in Cramp & Simmons (1993) to the effect that 'In times of stress, i.e. weather, [peregrines] may take carrion').

Rarer examples of avian predation come from Spain, where the plucked remains of an oriole were found in a pine wood near to where Goshawks nested, and from Switzerland, where birds migrating in autumn through the high Alpine passes in the canton of Valais run the gauntlet of hunting Eagle Owls. In a longitudinal study (1959–66) of prey taken by Eagle Owls, Salvan (1966b) lists an oriole among the numerous bird species taken. He notes that most of the hunting by this aggressive owl is nocturnal, and that birds are invariably taken in flight.

Illegal trapping and hunting

There are various reasons in different countries why orioles – and other species – are hunted: it may be simply for 'sport', or for food, or because the bird is regarded as a pest. In the former Chuvashian SSR, an infusion of oriole's feathers was considered a remedy against jaundice and malaria, and so the Golden Oriole was at one time an object of intensive shooting (Pershakov 1932). Moreover, information from the internet indicates that there is still a trade in bird skins; one Russian source quotes prices for many species, including orioles.

It is estimated that during spring passage, about 80,000 of that season's potential

Figure 47. In 1990, Channel 4 made a programme about the slaughter of migrant birds in the Mediterranean area. This photograph is the front cover of a booklet they produced to accompany the programme; it shows orioles hanging from a hunter's belt.

breeding birds are lost over the islands of the Maltese archipelago, where thousands of hunters kill birds for the sheer 'fun' of it. In addition, illegal trapping using nets, ground traps and liming still persists there and on many other Mediterranean islands. A number of the fatalities 'end up on the taxidermist's bench' (Fenech 1992). The figures given by Fenech suggest that the Golden Oriole is a prime target among passerines, being exceeded only by finch species at 2,000,000, Skylarks at 200,000 and Robins at 100,000. Evidence to give an idea of the number of orioles shot or otherwise killed on migration is provided by hunters' bag records and by taxidermists' records.

An analysis of 10 shooters' bag records of Golden Orioles killed showed an average of 4.93 per year and is illustrated in Table 54.

The records of three taxidermists in Malta are even more revealing and are detailed in Table 55.

When it is realised that in 1982 there were at least 40–50 taxidermists, the figures paint an even starker picture. Many of the birds are not even preserved in their own right, but are included as prey in the talons of a hawk.

Hopefully we can see an end to this now that Malta has joined the European Union and is subject to its laws on conservation. Under pressure from its hunters, the Maltese government negotiated derogation from the EU Birds Directive on

Shooter	*Years*	*No. of years*	*No. of Golden Orioles killed*	*Average (p.a.)*
A	1972–82	11	62	5.8
B	1982–87	6	54	9.0
C	1982–89	8	54	7.5
D	1982–86	5	30	6.0
E	1984–87	4	24	6.0
F	1984–87	4	15	3.75
G	1984–85	2	2	1.0
H	1984–87	3	18	6.0
I	1984–87	4	8	2.0
J	1984–87	4	9	2.25

Table 54. Analysis of 10 shooters' bags from Malta. Individual bags for one shooter, not included above, in spring 1986 showed 37 Golden Orioles taken, amongst other species.

joining in 2003, whereby springtime hunting for Common Quail and Turtle Dove was to be allowed between 25 March and 22 May. In effect, this also means that hunters will 'mistakenly' kill orioles that pass through at that time, thus giving a loophole to the shooting fraternity. BirdLife International is working on this matter and they point out that policing of the law is insufficient and have asked for it to be greatly improved. On Gozo Island in 2004, preventive measures were taken, and patrolling resulted in reducing much of the blatant illegal hunting and trapping, especially the use of mist nets (J. Mangion, Birdlife Malta President, *in litt.*).

On autumn migration, few orioles make landfall on Malta, but the killing does not stop, as many are shot and trapped on North African shores and along the Nile Valley. Hunters from Mediterranean countries now circumvent EU restrictions on hunting by arranging hunting trips to Egypt and North Africa.

Though Malta is probably the worst example, Italy has also had a bad reputation for trapping of passerines, mainly by netting and liming. Many of these small species are regarded as delicacies. The Lega Italiana Protezione Uccelli now believes that as a result of various publicity initiatives and a programme of education in schools, a climate of anti-hunting opinion is building up. Public markets for the

Taxidermist	*Years*	*No. of years*	*No. of Golden Oriole stuffed*	*Average (p.a.)*
A	1958–77	20	798	39.9
B & C	1976–86	11	228	20.7

Table 55. Maltese taxidermist records. Taxidermists B and C mounted birds for 35 and 30 clients respectively.

sale of birds are now banned, as is the restaurant trade serving 'polenta con osei' and other dishes containing small birds. In Greece, the situation has improved since it joined the EU, partly because Greek hunters are realising that to shoot small birds with expensive cartridges is not cost-effective (T. Akriotis, *in litt.*). Where it does continue, it is most often on remote, small islands like Antikythira and Gavdos. In Turkey and neighbouring Georgia, the trapping of raptors is still being practised, using such species as orioles, Red-backed Shrikes and Bee-eaters as lures. Hounding of migrating birds is practised to varying degrees in several other non-EU countries, but there is little evidence of orioles being targeted, except in Lebanon, where, despite the practice of liming having been made illegal, there are still reports of orioles being trapped in this way. Elsewhere some hunting still goes on in Lebanon, especially in the Bekaa Valley, mostly in spring, and in Georgia in autumn, where

Figure 48. That the hunting and trapping of orioles has been going on for millenia is illustrated by these Ancient Egyptian wall paintings. From Houlihan and Goodman, 1986; reproduced with kind permission of Aris & Philipps Publishers.

they are trapped and used as lure bait for trapping hawks and falcons to train for falconry.

In conclusion, one must acknowledge that the trapping of birds for sport or for the table has a long history. Paintings from the walls of various Pharaonic and other dignitaries of those times depict several instances of Golden Orioles. The paintings accurately portray the birds in both form and colour, even to the red bill. Baket III's tomb at Beni Hasan shows a flying male bird. Another depicts two orioles in a fig tree said to be engaged in 'playful contact'. This is from a trapping scene in a Userkaf 5th Dynasty (2465–2458 BC) mortuary temple at Saqqara. The mastaba of Niankhhunum and Khnumhotep (again 5th Dynasty) shows a Golden Oriole enmeshed in a net covering a fig tree. Another in the double tomb of a prominent official and his son (Akhethotep and Ptahotep), situated among a group of mastabas on the west side of the Step Pyramid at Saqqara, shows an oriole flying with a Eurasian Roller as they alight on the branch of a fig tree. Under the tree a man is kneeling as he sets a trap. There are similar scenes from other tombs showing flocks of Golden Orioles being netted as they feed on figs. Another scene depicts a flock of orioles in a tree of ripe figs and two men approaching the tree shouting and making other noises. This panics the birds, which fly blindly into a strategically placed, finely meshed net. The adjacent picture shows the birds being crated for market, where they were to be sold for the table. The next illustration is of trappers plucking the birds. Inscriptions in other tombs imply that stick throwing was also employed, possibly by casual hunters. Oriole flesh is said to be good, as its taste will have been improved by its feeding on so much sweet fruit. The practice continues to this day; orioles were seen being trapped in the spring of 1984 near Port Said (Houlihan & Goodman 1986).

CHAPTER FIFTEEN

Wintering

After migrating south along three main routes, the birds arrive in their wintering area, which includes Kenya (mostly the south), Uganda, Zambia, Zimbabwe, Tanzania, Mozambique, Angola, Cameroon, eastern parts of DR Congo, Botswana and down the eastern and central parts of South Africa, in both the Eastern and Western Cape provinces (McGill and Nixon 1991). There are also some occasional records from Madagascar, mostly during migration periods. It was at one time thought that there could be a smaller area in West Africa where a discrete population originating from Morocco and Iberia wintered (Feige 1986). This belief was based on a few sightings but it is now thought that these birds do not constitute a different population, although there are still the occasional records from countries like Equatorial Guinea. It was Moreau (1961) who initiated the theory that wintering occurred somewhere between Equatorial Guinea and Gambia. This is in contrast to the hypothesis that the birds fly over the Sahara desert southward in a diagonal direction (Stresemann 1948). There is very little evidence now for this latter belief. There is now more evidence that, having made the southward journey

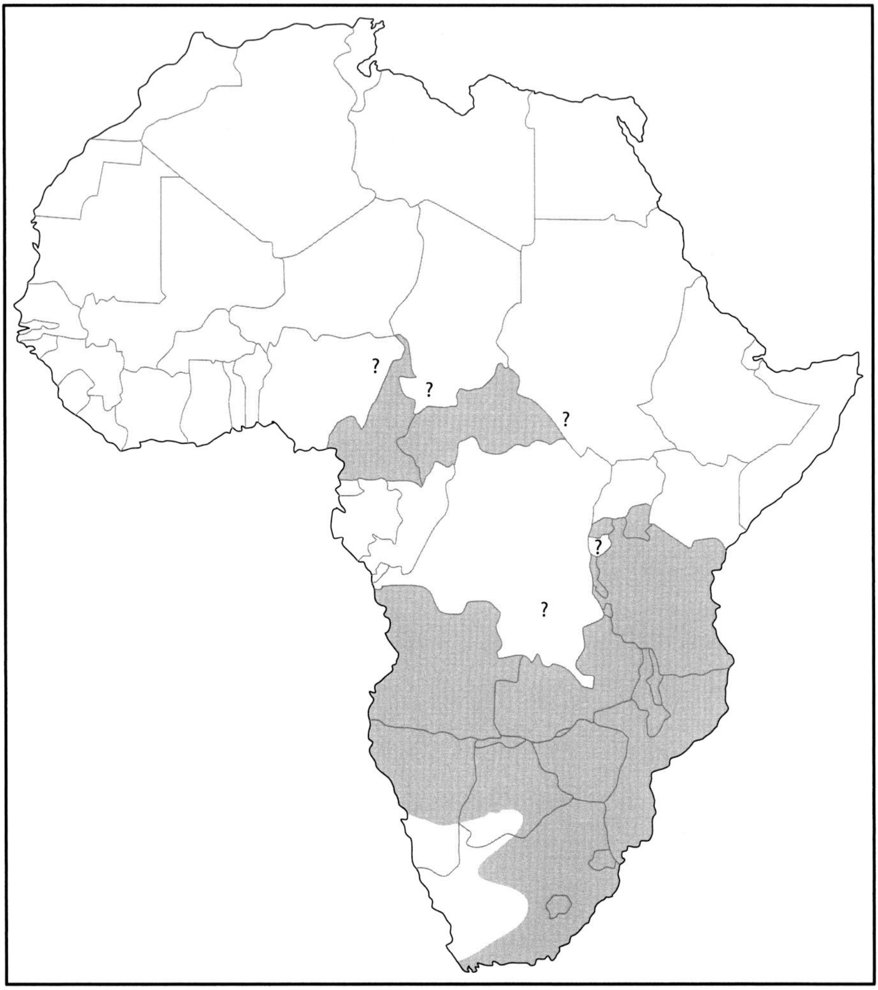

Figure 49. Main wintering areas of Golden Orioles in Africa. The question marks indicate unconfirmed data. After Baumann, 2001.

at least as far as Mauritania, where many Eurasian Golden Orioles have been seen and ringed (V. Salewski, pers. comm.), the birds pass on to the sub-Saharan savanna. Only two verifiable records exist from Guinea and these were in early November. Peter Jones (*in litt.*), an expert on West African ornithology, says he has not seen Golden Orioles in midwinter in the region, although they are frequent on passage. An alleged midwinter sighting of a female/young in Ghana was dismissed because of possible confusion with African Golden Oriole *O. auratus*, and there are no reliable wintering records from Nigeria. If it winters in West Africa at all, it will almost certainly be in the Guinea savanna, a deciduous, broadleaved woodland of (amongst others) *Isoberlinia* spp. that is ecologically similar to miombo. Guinea savanna extends in a broad latitudinal belt right across West and central Africa, just south of the Sudan vegetation zone and north of the rainforest. It is quite possible the orioles all head south of the equator from West Africa – a young one was found dead well off course in mid-November 2003 on St Helena, nearly 2,000km out in the southern Atlantic Ocean (Philip Ashmole, *in litt.*). Whether they cross over the rainforest to make for Congo, Angola and South Africa is a moot point. The island of São Tomé does have Eurasian Golden Oriole on its species list as a migrant, but authenticated records and dates are hard to come by. What seems to be more likely is that the migrating birds follow the savanna to Cameroon and then move south and south-east.

From most of the rest of Europe the birds take the route along the River Nile via Sudan to Kenya and the area around Lake Victoria, having been joined by others that had come across the Red Sea and along the Rift Valley. Sabine Baumann (1999) examined over 5,000 records and found that the winter distribution, defined as being between December and February, was divided into two areas: (1) the rainforests of the Central African Republic, principally below 5° S, and (2) a less significant area in the north of Cameroon. Her research also showed that the birds are found in a variety of habitats, including riverine and gallery forest types and forest savannas. There are very few records from the tropical rainforests themselves.

An analysis of figures supplied by Bruno Walther (*in litt.*) of records of sightings shows that while on migration birds can reach considerable elevations, from 2,300m in Rwanda to as high as 4,300m, where a bird was found dead on Kilimanjaro in Tanzania (Sonnenschein 1998). For most of the recognised wintering months, November–February, they are found at much lower altitudes, from 900m (850m in Angola) down to 50m in Kenya.

Whereas most European passerine migrants in Africa winter in the semi-arid savannah just south of the Sahara, the Eurasian Golden Oriole is one of only a few species that move further south into moister deciduous wooded habitat found in the south and south-east of Africa, typically into mopane and miombo woodland (Baumann 1999). It can also be found at times in some non-indigenous plantations and in gardens and orchards, but is most often observed in patches of tall trees of mixed species.

A wintering Golden Oriole usually moves solitarily, but several birds will gather at fruiting trees such as figs. They also seem to follow behind moving thunderstorms and similar wet weather patterns. Many observers over the years have noted that Eurasian Golden Orioles have been found alongside the endemic orioles, but not until Sabine Baumann (2000b) carried out her research in Zimbabwe had anybody studied interspecific relations in these orioles. She found that in her Zimbabwean study area the African Black-headed Oriole, *O. larvatus*, is the commonest oriole, found in moist woodland, both montane forest and miombo. The African Golden Oriole *O. auratus* seems to prefer miombo, where it outnumbers the other oriole species, although it is restricted to the taller broadleaved

Figure 50. Miombo habitat in Zimbabwe – typical oriole wintering habitat. S. Biechle.

trees. *Oriolus oriolus* arrives in the area between the last days of September and early December and remains until March. *O. auratus* is a partial migrant, but is always present as a breeding bird during the rainy season, i.e. from September to January, after which some will leave south-east Africa and move to East Africa. *O. larvatus* also breeds between the months of September and February, the peak breeding season being October–November, but, unlike *O. auratus*, it is sedentary.

Baumann carried out her reserach over a period of six weeks in February and March, when, she says, all three species were 'abundant'. After assessing the habitat, she defined twenty-one 700m^2 'bird-use' plots where *O. oriolus* fed consistently for over 10 minutes at a time. Most kinds of habitat were represented in the plots, including miombo/mopane, acacia, gardens and riverine areas. Then she did the same for the other two species. The next step was to identify which of the 21 'bird-use' plots were used by all three species for feeding. She observed their interactions, behaviour and foraging and compared them with other plots that did not contain all three species.

She then took note of the height of the trees used most by the birds and watched each bird for periods of 60–90 seconds at a time when they were feeding. The observations included:

1. Foraging movements: perched, hanging, jumping, fluttering and hovering
2. Song: although *O. oriolus* sing in Africa they do not defend territories there. These songs were recorded for playback later and used 30m away to gain a

Figure 51. Mopane habitat in Zimbabwe. Sabine Baumann.

reaction from all three species. This was done for each of the other two species in turn.

Summing up her interpretation of the results of the experiment she says the *O. oriolus* were mostly in pairs or groups of three to four birds. They were quite vocal when in the company of the other African orioles. However, *O. oriolus* and *O. auratus* songs are remarkably similar. The response between these two was greater than that between *O. oriolus* and *O. larvatus*. Whilst foraging, there was more antagonism between *O. auratus* and *O. oriolus*, but little if any from *O. larvatus* to *O. oriolus*. As regards reactions in nesting trees, she found that the heights at which *O. larvatus* and *O. auratus* nested differed appreciably. *O. auratus* nested on average at 6.5–7.0m high wheras *O. larvatus* stayed at about 5.3m. Most of the *O. auratus* nests were in *Brachystegia* species but *O. larvatus* was much more catholic as regards tree species.

The population of *O. auratus* was greater than the sum of the other two added together in her study area, but it has to be remembered that *O. oriolus* is nomadic during this time. Because the winter visitors remained mainly in the top section of any tree canopy and *O. auratus* was mainly in the centre section, there was no significant physical antagonism between the two. Because *O. larvatus* was in the lower section antagonism towards *O. oriolus* was virtually nil, as the former foraged inside the canopy on larger branches and on the trunks, in the manner of woodpeckers, while the latter stayed more on the outside of the canopy at a greater

	O. oriolus	*O. larvatus*
Number of items	59	41
% insects*	25	37
% insects taken off leaves	93	80
% food items caught in air	12	24
Feeding height (mean + SD)	8.8, SD 0.8	6.8, SD 1.0

Table 56. Feeding behaviour of orioles in the Tsavo, Kenya. *All other food items were fruit. From Lack, 1985.

height. As regards food items, all the orioles fed on similar prey items, such as hairy. After the the endemic orioles have finished breeding, *O. oriolus* can often be seen travelling in company with *O. auratus*.

In his notes on avian ecology in the Tsavo area of Kenya, Lack (1985) records that the same three species were present, but he recorded feeding behaviour of only two, *O. larvatus* and *O. oriolus* (Table 56). The difference between the two species is not statistically significant. It does, however, provide a further confirmation of Baumann's findings that *O. larvatus* feed in, and occupy, lower strata than *O. oriolus*, thereby reducing competition.

Further evidence of the Eurasian Golden Oriole's wintering range comes from ringing studies carried out in Kenya, Uganda and Angola, involving a total of 431 birds (Kenya, 384 – three different ringers; Uganda 42 – two different ringers; Angola 5 birds – one ringer). There have been no recoveries to date. Table 57 shows the duration and distribution of these ringing efforts and the period during which the birds were ringed. The Ugandan birds at Bwindi Impenetrable Forest and at the River Kafu Palm Savanna give an indication of the wide range of habitat used.

Country/period	*No. ringed*	*Ringing dates*	*Location*
Kenya 1971–99	8	'winter dates'	various
	366	Sept–Apr	various
	10	Oct–Dec	north of Nakuru
Uganda 1969–71	24	Nov	Mbabara/Kampala
	16	Nov–Mar	not specified
	1	2 May	Bwindi
	1	10 Aug	River Kafu
Angola 1968–99	4	Nov–Feb	various
	1	15 Sept	not specified

Table 57. Ringing dates and distribution in Kenya, Uganda and Angola.

CHAPTER SIXTEEN

Orioles in captivity

Over the Christmas period of 2005 a Golden Oriole turned up at Leighton Moss RSPB reserve in Lancashire (Lythe 2006). The same bird had also been reported from nearby earlier in the month. Not only is this place in north-west England but the weather was cold, windy and snowy and most unlike the situation where one would expect to find a bird that should be enjoying the sun of Africa. The surprise might have been less if the bird had been a juvenile or immature, which are known to 'reverse migrate' at times. Even records for these at this time of year are unprecedented, the hangers-on usually being recorded no later than November. A record at such a date in southern Europe, as occurred in the case of a ringing recovery from Greece on 4 December 1990, would be less surprising. The point about the Lancashire bird was that it was a full-adult yellow male, an occurrence so unusual that it was even chronicled in the press. The question arose as to the likelihood that it was an escaped cage bird. It was seen to feed on the ground at times, which is uncommon but not unknown in the wild, but is much more likely to be part of the behaviour of captive bird.

In many Far East Asian countries, species of colourful singing birds are kept in cages. The birds are traded in markets in Thailand, Malaysia and Indonesia, among others. At Medan market in Sumatra, the Black-naped Oriole is described as a species very often bought and sold. There is also one record of Black-naped Oriole being bred in captivity (Schiffer 2005). There is also a 1965 record in an article in Avicultural Magazine of a pair of African Black-headed Orioles being kept in the United States. They had been captured in Mozambique in 1960 and exported, though the male of the pair was killed by a Green Cardinal in transit. A replacement male was sent in 1962 and successful breeding occurred in 1964.

After an initial claim that no Golden Orioles had been bred in captivity (Kramer 1980), a German breeder, Hermann Greulich in Rauenberg, successfully reared three young and two young in successive years (Graff 1990, 1991), although it is clear from Greulich's account that it was no easy task and required constant attention from him. Another breeder, Rudiger Neff (2003), again in Germany, has kept the species since 1982. Perhaps the most interesting information to come out of it all is that one male had lived, by that time, to 18 years old and was still fertile. There is a case of even greater longevity; a bird owned by another person lived 20 years. Suggestions have been made that they may even reach 25 years.

A puzzling incident occurred in the fenland study area when a complete nest disappeared when the eggs were known to be near to hatching. The forked branch where the nest was had been neatly cut or sawn off. It is unlikely that egg thieves would take a whole nest: apart from anything else, it exposes them for a longer time to discovery. We suspect that this nest was taken by a cage-bird enthusiast so that he could continue to rear the chicks after hatching, and although a local person was suspected, it proved impossible to take the matter further.

EU and other laws forbid the trapping of many species, including orioles, for cage-bird purposes but they do allow for birds 'reared in captivity' to continue to be kept, the inference being that any breeders must, by now, have kept them for a very long time and before the present laws were enacted. Meaden (1979) records that Golden Orioles are kept and goes on to say that he had only ever had one pair that he had obtained from 'abroad'. He kept them in a 5m-long, 3m-high and 3m-wide enclosure and fed them on soft food, consisting of vegetable matter with soft fruit, together with locusts as well as mealworms and even small fish. He comments 'a real effort must be made to breed from them to even justify having them', which supports the view that they are difficult birds to breed in captivity. The food given was basically man-made. All the evidence is that Golden Orioles are not easy to rear in captivity, so the provenance of the Leighton Moss bird remains a mystery.

APPENDIX 1

Systematic list of Oriolidae

The following listing of the genera *Oriolus* and *Sphecotheres* is based on Howard and Moore, *A Complete Checklist of the Birds of the World*, (2003), updated with reference to Sibley & Monroe (1993) and, more recently, del Hoyo *et al.* in *Handbook of Birds of the World* (Vol 13).

ORIOLIDAE

ORIOLUS

The 'mimeta' group

Brown Oriole *Oriolus szalayi*
New Guinea, W. Papuan Is
Dusky-brown or Moluccan Oriole *Oriolus phaeochromus*
Halmahera Is (N. Moluccas)
Grey-collared Oriole *Oriolus forsteni*
Ceram Is
Black-eared Oriole *Oriolus bouroensis*
O. b. bouroensis
Buru Is S. Moluccas
O. b. decipiens
Tanimbar Is (Lesser Sundas)
Olive-brown Oriole *Oriolus melanotis*
O. m. melanotis
Timor, Roti, Seman (Lesser Sundas)
O. m. finschi
Wetar Is (Lesser Sundas)
Olive-backed Oriole *Oriolus sagittatus*
O. s. magnirostrus
SC New Guinea
O. s. affinis
NW & NC Australia (S. Gulf of Carpentaria)

O. s. grisescens
N. Queensland, Cape York
O. s. sagittatus
NE & SE Australia

Green Oriole or Australian Yellow Oriole *Oriolus flavocinctus*
O. f. migrator
Leti, Moa and Roma Islands (E. Lesser Sundas)
O. f. muelleri
Aru Islands and SC New Guinea
O. f. flavocinctus
N. Western Australia (Kimberley) and Northern Territory (Arnhem Land)
O. f. tiwi
Melville and Bathurst Islands (Northern Territory)
O. f. flavotinctus
N Queensland/Cape York
O. f. king
NE Queensland.

Dark-throated Oriole *Oriolus xanthonotus*
O. x. xanthonotus
S. Burma, Thailand, Malay Peninsula, Sumatra, Java, Bangka Islands, SW Borneo.
O. x. mentawi
Siberut, Sipura, Pagai (Mentawai Islands, W. Sumatra).
O. x. consobrinus
N., C. & E. Borneo and adjacent islands
O. x. persuasus
Palawan and Culion Islands (Philippines)

Philippine Oriole *Oriolus steerii*
O. s. albiloris (White-lored Oriole)
N. Luzon.
O. s. samarensis
Samar, Leyte, Bohol, E. Mindanao
O. s. assimilis
Cebu (last recorded 1906 but unconfirmed sighting in 2000)
O. s. steerii
Masbate, Negros
O. s. basilanicus
Basilan, W. Mindanao
O. s. cinereogenys
Tawitawi, Bongao and possibly Baliungan (Sulu Archipelago)

Isabela Oriole *Oriolus isabellae*
Mountains of Bataan and Isabela Provinces, Luzon

The 'yellow-headed' group

Eurasian Golden Oriole *Oriolus oriolus*
Europe and N. Africa eastwards to W. Siberia south of *c.* 60° N to Iran. Winters Africa and NW India.

Indian Golden Oriole *Oriolus kundoo*
Uzbekistan, S. Kazakhstan, Afghanistan, Tien Shan and Indian Peninsula, south to Karnataka and N. Tamil Nadu. Winters further south on Indian Peninsula

African Golden Oriole *Oriolus auratus*

O. a. auratus
W. & NC Africa

O. a. notatus
Eastern & Southern Africa

Slender-billed Oriole *Oriolus tenuirostris*

O. t. tenuirostris
Himalayan foothills from E. Nepal, Assam, Bhutan to C. Burma, SW Yunnan, N. Laos winters S. Burma, W. and SW Thailand.

O. t. invisus
Vietnam

Black-naped Oriole *Oriolus chinensis*

O. c. diffusus (Eastern Black-naped Oriole)
E Asia: throughout China from W. Sichuan to Manchuria and Amurland, Yunnan, N. Laos, Hainan and Taiwan winters S. India and Sri Lanka, Malaysia, Indochina

O. c. andamanensis (Andaman Black-naped Oriole)
Andaman Islands

O. c. macrourus (Nicobar Black-naped Oriole)
Nicobar Islands

O. c. maculatus
Singapore and south Malay Peninsula, mainland Sumatra, Java, Borneo, Bali, Bangka and Belitung Islands.

O. c. mundus
Simeulue and possibly Nias Islands, W. Sumatra

O. c. sipora
Sipura Island, W. Sumatra

O. c. richmondi
Siberut and Pagai Islands, W. Sumatra

O. c. lamprochryseus
Masalembu and Keramian Islands, Java Sea

O. c. insularis
Kangean, Sapudi and Raas Islands, Bali Sea, Indonesia

O. c. melanisticus
Kaburuang, Karakelong and Salibabu (Talaut Islands), Indonesia

O. c. sangirensis
Sangirand Tabukan Islands (N. Sangir archipelago), Indonesia
O. c. formosus
Siau, Ruang, Tahulandang, Biaro Islands (S. Sangir archipelago) and Mayu Island, Indonesia
O. c. celebensis
Sulawesi and surrounding islands
O. c. stresemanni
Peleng (Banggai archipelago), Indonesia
O. c. frontalis
Taliabu (Sula archipelago)
O. c. boneratensis
Flores Sea Islands, Tukangbesi islands, Banda Sea, Indonesia
O. c. broderipi
Lombok, Sumbawa, Komodo, Sumba, Flores, Besar, Rinca, Paloe, Lomblen, Pantar, Alor Islands
O. c. chinensis
NW Philippines
O. c. yamamurae
SE Philippines
O. c. suluensis
Sulu archipelago, Philippines

African 'black-headed' group

Green-headed Oriole *Oriolus chlorocephalus*
O. c. amani
SE Kenya, NE to SE Tanzania
O. c. chlorocephalus
S. Malawi and C. Mozambique (Mt Chiperone)
O. c. speculifer
S. Mozambique (Mt Gorongoza)
São Tomé or Great-billed Oriole *Oriolus crassirostris*
São Tomé Is
Western Black-headed Oriole *Oriolus brachyrhynchus*
O. b. brachyrhynchus
W. Africa
O. b. laetior
WC & C. Africa
Abyssinian, Dark-headed or Forest Oriole *Oriolus monacha*
O. m. monacha
Eritrea, N. Ethiopia

O. m. meneliki
S. Ethiopia
Mountain or Black-tailed Oriole *Oriolus percivali*
Zaire, Kenya
African Black-headed Oriole *Oriolus larvatus*
O. l. rollet
S. Sudan, Ethiopia, Uganda, Kenya, Lake Victoria basin and E. DR Congo
O. l. reichenowi
Somalia southwards to E. Tanzania
O. l. angolensis
Angola, N. Namibia to N. Mozambique and W. Tanzania
O. l. additus
Coastal S. Tanzania to coastal S. Mozambique
O. l. larvatus
S. Zimbabwe, S. Mozambique, E. South Africa
Oriolus nigripennis (Black-winged Oriole)
O. n. alleni
W. Africa
O. n. nigripennis
WC & C. Africa

Asian 'black-headed' group

Black-hooded or Asian Black-headed Oriole *Oriolus xanthornus*
O. x. xanthornus
N. India, Burma, Thailand, Indochina, N. Malaysia winters Sumatra
O. x. maderaspatanus
Ganges plain and peninsular India
O. x. ceylonensis (Sri Lanka Black-headed Oriole)
Sri Lanka
O. x. reubeni
Andaman Islands
O. x. tanakae
NE Borneo and adjacent islands
Black Oriole *Oriolus hosii*
Borneo
Black-and-crimson or Crimson-breasted Oriole *Oriolus cruentus*
O. c. malayanus
Peninsular Malaysia
O. c. consanguineus
Mainland Sumatra
O. c. cruentus
Java

O. c. vulneratus
N. Borneo
Maroon Oriole *Oriolus traillii*
O. t. traillii
Himalayas, Burma, SW China, N. Thailand, N. Laos, N. Vietnam
O. t. robinsoni
S. Laos and S. Vietnam
O. t. nigellicauda
Hainan Island. Winters N. Indochina
O. t. ardens
Taiwan
Silver Oriole *Oriolus mellianus*
W. China

FIGBIRDS

Timor Figbird *Sphecotheres viridis*
Timor and Roti Islands (Lesser Sundas)
Wetar Figbird *Sphecotheres hypoleucus*
Wetar Island (Lesser Sundas), Indonesia
Green Figbird or **Yellow Figbird** *Sphecotheres vieilloti*
S. v. salvadorii
SE New Guinea
S. v. cucullatus
Kai Is
S. v. ashbyi
N. Western Australia and N. Northern Territory
S. v. flaviventris (Yellow Figbird)
Cape York and NE Queensland
S. v. vieilloti (Green Figbird)
E. Australia

APPENDIX 2

List of species mentioned in the text

BIRDS

Barn Owl	*Tyto alba*
Barn Swallow	*Hirundo rustica*
Bearded Tit	*Panurus biarmicus*
Black Drongo	*Dicrurus adsimilis*
Black Kite	*Milvus migrans*
Black Sparrowhawk	*Accipiter melanoleucus*
Black Woodpecker	*Dryocopus martinus*
Black-billed Cuckoo	*Coccyzuz erythrophthalmus*
Blackcap	*Sylvia atricapilla*
Black-faced Friarbird	*Philemon moluccensis*
Black-headed Gull	*Larus ridibundus*
Black-vented Oriole	*Icterus wagleri*
Blue Tit	*Parus caeruleus*
Bobolink	*Dolichonyx orizyvorus*
Bullfinch	*Pyrrhula pyrrhula*
Carib Grackle	*Quiscalus lugubris*
Carrion Crow	*Corvus corone*
Cetti's Warbler	*Cettia cettia*
Chaffinch	*Fringilla coelebs*
Chiffchaff	*Phylloscopus collybita*
Collared Dove	*Streptopelia decaocto*
Collared Flycatcher	*Ficedula albicollis*
Common Blackbird	*Turdus merula*
Common Buzzard	*Buteo buteo*
Common Cuckoo	*Cuculus canorus*
Common Magpie	*Pica pica*
Common Nightingale	*Luscinia megarhynchos*
Common Quail	*Coturnix coturnix*
Common Redstart	*Phoenicurus phoenicurus*
Common Snipe	*Gallinago gallinago*
Common Whitethroat	*Sylvia communis*

Common Woodpigeon	*Columba palumbus*
Corn Bunting	*Emberiza calandra*
Dalmatian Pelican	*Pelicanus crispus*
Dartford Warbler	*Sylvia undata*
Dunnock	*Prunella modularis*
Dusky Friarbird	*Philemon fuscicapillus*
Eagle Owl	*Bubo bubo*
Eleonora's Falcon	*Falco eleonorae*
Eurasian Jay	*Garrulus glandarius*
Eurasian Kestrel	*Falco tinnunculus*
Eurasian Robin	*Erithacus rubecula*
Eurasian Roller	*Coracius garrulus*
Eurasian Starling	*Sturnus vulgaris*
Eurasian Treecreeper	*Certhia familaris*
European Bee-eater	*Merops apiaster*
Fieldfare	*Turdus pilaris*
Garden Warbler	*Sylvia borin*
Goshawk	*Accipiter gentilis*
Great Bittern	*Botaurus stellaris*
Great Spotted Woodpecker	*Dendrocopos major*
Great Tit	*Parus major*
Green Woodpecker	*Picus viridis*
Grey Heron	*Ardea cinerea*
Grey-headed Woodpecker	*Picus canus*
Grey-necked Friarbird	*Philemon subcorniculatus*
Hawfinch	*Coccothraustes coccothraustes*
Helmeted Friarbird	*Philemon buceroides*
Hen Harrier	*Circus cyaneus*
Hobby	*Falco subbuteo*
Honey Buzzard	*Pernis apivorus*
Hooded Crow	*Corvus cornyx*
Hoopoe	*Upopa epops*
House Sparrow	*Passer domesticus*
Icterine Warbler	*Hippolais icterina*
Kingfisher	*Alcedo atthis*
Lesser Whitethroat	*Sylvia curruca*
Levaillant's Green Woodpecker	*Picus valliantii*
Little Egret	*Egretta garzetta*
Long-eared Owl	*Asio otus*
Madagascar Bee-eater	*Merops superciliosus*
Mallard	*Anas platyrhynchos*
Marsh Harrier	*Circus aeruginosus*
Martinique Oriole	*Icterus bonana*

Meadow Pipit	*Anthus trivialis*
Mediterranean Gull	*Larus melanocephalus*
Metallic Starling	*Aplonis metallica*
Middle Spotted Woodpecker	*Dendrocopos medius*
Mistle Thrush	*Turdus viscivorus*
Montserrat Oriole	*Icterus oberi*
Mourning Dove	*Zenaida macroura*
Northern Oriole	*Icterus galbula*
Northern Wheatear	*Oenanthe oenanthe*
Peregrine Falcon	*Falco peregrinus*
Pheasant	*Phasianus colchicus*
Pied Flycatcher	*Ficedula hypoleuca*
Puffin	*Fratercula arctica*
Purple Heron	*Ardea purpurea*
Red Kite	*Milvus milvus*
Red-backed Shrike	*Lanius collurio*
Redshank	*Tringa totanus*
Red-shouldered Hawk	*Buteo lineatus*
Reed Bunting	*Emberiza schoeniclus*
Reed Warbler	*Acrocephalus scirpaceus*
River Warbler	*Locustella fluviatilis*
Rook	*Corvus frugilegus*
Shiny Cowbird	*Molothrus bonariensis*
Skylark	*Alauda arvensis*
Song Thrush	*Turdus philomelos*
Sooty Falcon	*Falco concolor*
Spanish Sparrow	*Passer hispaniola*
Sparrowhawk	*Accipiter nisus*
Spotted Flycatcher	*Muscipapa striata*
Streak-headed Honeyeater	*Pycnopygius stictocephalus*
Tawny Owl	*Strix aluca*
Tree Sparrow	*Passer montanus*
Tropical Bulbul	*Laniarius aethiopicus*
Turtle Dove	*Streptopelia turtur*
White-tailed Eagle	*Heliaeetus albicilla*
Wood Lark	*Lullula arborea*
Wood Warbler	*Phylloscopus sibilatrix*
Wren	*Troglodytes troglodytes*
Wryneck	*Jynx torquilla*
Yellow Wagtail	*Motacilla flava*
Yellow-breasted Chat	*Icteria virens*
Yellowhammer	*Emberiza citrinella*

MAMMALS

Beaver	*Castor fiber*
Ferret	*Mustela putorius*
Grey Squirrel	*Neosciurus carolinensis*
Red Squirrel	*Sciurus vulgaris*
Short-tailed Field Vole	*Microtus agrestis*

INSECTS

Moths and butterflies

Black-veined White	*Aporia crataegi*
Brown-tailed Moth	*Euproctis chrysorrhoea*
Buff-tip	*Phalera bucephala*
Convolvulus Hawkmoth	*Agruis convolvuli*
Emperor Moth	*Saturnia pavonia*
Engrailed	*Ectropis bistortata*
European Oak Leafroller	*Tortrix viridana*
Eyed Hawkmoth	*Smerinthus ocellata*
Garden Tiger	*Arctia caja*
Giant Leopard Moth	*Hypercompe scribonia*
Goat Moth	*Cossus cossus*
Great Prominent	*Peridea anceps*
Green Oak Tortrix	*Tortrix viridana*
Gypsy Moth	*Lymantria dispar*
Herald Moth	*Scoliopterix libatrix*
Hornet Clearwing Moth	*Sesia apiformis*
Lappet Moth	*Gastropacha populifolia*
Large Emerald	*Geometra papilionaria*
Large White	*Pieris brassica*
Leopard Moth	*Zeuzera pyrina*
Lesser Emperor	*Anax parthenope*
Lime Hawkmoth	*Mimas tiliae*
Mopane Emperor Moth	*Imbrasia belina*
Northern Winter Moth	*Operophtera fagata*
Oak Eggar	*Lasiocampa quercus*
Painted Lady	*Vanessa cardui*
Pine Hawkmoth	*Hyloicus pinastri*
Pine-tree Lappet Moth	*Dendrolimus pini*
Poplar Hawkmoth	*Laothoe populi*
Privet Hawkmoth	*Sphynx ligustri*
Silkmoth	*Bombyx mori*

Silver-Y Moth	*Autographa gamma*
Small Tortoiseshell	*Aglais urticae*
Swallow Prominent	*Pheosia tremula*
Sycamore	*Acronicta aceris*
Lackey Moth	*Malacosoma neustria*
Wheat Chafer	*Anisoplia austriaca*
White Satin Moth	*Leucoma salicis*

OTHER INSECTS

Birch Beetle-sawyer	*Cimbex femorata*
Cicada	*Platypleura rutherfordi*
Cockchafer	*Melolontha melolontha*
Common Grasshopper	*Tettigonia caudata*
European Garden Spider	*Araneus diadematus*
Great Green Bush Cricket	*Tettigonia viridissima*
Green Grasshopper	*Tettigonia caudata*
Green Shield Bug	*Palomena prasina*
Honey Bee	*Apis mellifera*
Locust	*Locusta migratoria*
Mealworm	*Tenebrio molitor*
Migrant Hawker	*Aeshna mixta*
Small Red-eyed Damselfly	*Ceriagrion tenellum*
Yellow Garden Ant	*Lasius flavus*

FLORA

Acacia	*Acacia* spp.
African Mahogany	*Khaya nyasica*
Alder, Black	*Alnus glutinosa*
Almond	*Prunus dulcis*
Apple	*Malus domesticus*
Ash, European	*Fraxinus excelsior*
Australian Flame Tree	*Brachychiton acerfolia*
Birch, Downy	*Betula pubescens*
Birch, Siberian Silver	*Betula platyphylla*
Birch, Sichuan Birch	*Betula szechuanica*
Birch, Silver	*Betula pendula*
Birch, White	*Betula verrucosa*
Bird Cherry	*Prunus padus*
Bloody-red Hawthorn	*Crataegus sanguinea*
Broad-leaved Erythrina	*Erythrina latissima*

Castor Oil	*Ricinus communis*
Cherry, Wild	*Prunus avium*
Christ's Thorn	*Paliurus aculeatus*
Common Reed	*Phragmites australis*
Coral Tree	*Erythnia* spp.
Cork Oak	*Quercus Suber*
Dzhidda	*Elaeagnus oxycarpa*
Elder	*Sambucus nigra*
Elm, English	*Ulmus procera*
Elm, Wych	*Ulmus glabra*
European Aspen	*Populus tremula*
False Acacia	*Robinia pseudoacacia*
Forest Fig	*Ficus craterostoma*
Green Bristle Grass	*Setaria visidis*
Hemlock	*Conium maculatum*
Holm Oak	*Quercus ilex*
Hornbeam	*Carpinus betulus*
Judas Tree	*Cercis siliquastrum*
Karagach	*Ulmus campestris*
Larch	*Larix siberica*
Loquat	*Eriobotrya japonica*
Mowah Tree	*Bossia* spp.
Mulberry, Black	*Morus nigra*
Mulberry, White	*Morus alba*
Myrtle, Common	*Myrtus communis*
Neem tree	*Azadirachta indica*
Olive	*Olea europea*
Oriental Sweet Gum	*Liquidambar orientalis*
Pampas Grass	*Cortaderia* spp.
Pea Tree	*Caragana* spp.
Peach	*Persica vulgaris*
Pear	Genus *Pyrus*
Pea Tree or Karagana	*Caragana arborescens*
Pine, Scots	*Pinus sylvestris*
Pine, Stone	*Pinus pinea*
Pistachio	*Pistacia lentiscus*
Plane	*Platanus orientalis*
Raspberry	*Rubus idaeus*
Rosemary	*Rosmarinus officinalis*
Sea Buckthorn	*Hippophae rhamnoides*
Shepherd's Tree	*Boschia albitrunca*
Silk Cotton	*Bombax* spp.
Silky Oak	*Greyvilia* spp.
Silver Birch	*Betula pendula*

Small-leaved Lime	*Tilia cordata*
Spruce, Norway	*Picea abies*
Sugar Beet	*Beta vulgaris* var. *saccharifera*
Sweet Chestnut	*Castanea sativa*
Sycamore Fig	*Ficus sycomorus*
Tamarind	*Tamarindus indica*
Tamarisk	Genus *Tamarix*
Transvaal Saffron	*Cassine transvaalensis*
Turpentine Tree	*Colophospermum mopane*
Usnea	*Tillandsia usneoides*
Walnut	*Juglans* sp.
Wild Plum	*Prunus sogdiana*
Crack Willow	*Salix fragilis*
Weeping Willow	*Salix babylonica*
Xen Oak	*Quercus canarensis*

APPENDIX 3

Invertebrate prey

Summary of invertebrate prey items identified in Foggo (1993)

Pemphigidae spp.	poplar spiral gall aphid
Aphidae spp. (unidentified)	aphids
Chaetophoridae spp.	aphids
Philaenus spumarius	frog hopper
Alnetoidea alneti	plant hopper
Cicadellidae (unidentified)	plant hopper
Macropsis scutellata	
Anthocoris nemorum	small bug
Microphysidae (unidentified)	minute bugs
Miridae (unidentified)	capsid bugs
Nabidae (unidentified)	damsel bugs
Agromyzidae	leaf-mining bugs
Anisopodidae	–
Asilidae	robber flies
Contarinia petioli	poplar gall midge
Ceratopogonidae	biting midges
Chronomidae	(biteless) midges
Chiromyidae	–
Culcidae	mosquitoes
Chlorops hypostigma	–
Dolichopodidae	long-headed flies
Empididae	–
Cnemcantha spp.	–
Lonchaea chorea	–
Micropezidae	stilt-legged flies
Milichiidae	–
Mycetophilidae	fungus gnats
Toxoneura muliebris	–
Phoridae	–
Pipunculidae	–
Platypezidae	–
Psychodidae	–

Rhagio scolopacea	
Sepsidae	ant-like flies
Pachygaster atra	soldier flies
Syrphidae	hover flies
Syrphidae larvae	hover fly larvae
Cantharis fulvicollis	soldier beetles
Rhagonycha fulva	soldier beetles
Cantharis nigricans	soldier beetles
Cantharis rufa	soldier beetles
Psilothrix cyaneus	soldier beetles
Dromius quadrinotatus	ground beetles
Dromius quadrimaculatus	ground beetles
Carabidae (unidentified)	ground beetles
Dromius linearis	ground beetles
Dromius melanocephalus	ground beetles
Poecilium alni	longhorn beetles
Phratora laticollis	willow leaf beetle
Chalcoides aurata	willow leaf beetle
Longitarsus spp.	flea beetles
Chryomelid larvae	leaf beetles
Clavicornids (unidentified)	small beetles
Adalia bipunctata	two-spot ladybird
Adalia decempunctata	ten-spot ladybird
Coccinellid larvae	ladybird larvae
Cholevidae	small beetles
Apion carduorum	weevils
Phyllobius parvulus	weevils
Apion nigritarse	weevils
Dorytomus longimanus	weevils
Dorytomus validirostris	weevils
Sciaphilus asperatus	weevils
Phyllobius viridearis	weevils
Rhampus pulicarius	weevils
Apion spp.	weevils
Balanius spp.	weevils
Otiorrhynchus spp.	weevils
Adrastus spp.	click beetles
Athous haemorrhoidalis	click beetles
Dolopius marginatus	click beetles
Gonodera murina	–
Gonodera luperus	–
Laria rufipes	–
Coleop larvae (unidentified)	–
Ptilinus pectinicornis	fan-bearing boring beetle

Rhinosimus planirostris	–
Scolytidae (unidentified)	bark beetles
Apocrita	bees
Formicida	ants
Symphyta	sawflies
Parasitica	parasitic wasps
Notodontidae (unidentified)	large caterpillars
Furcula bifida	poplar kitten moth
Laothoe populi	poplar hawkmoth
Smerinthus ocellata	eyed hawkmoth
Poecilocampa populi	december moth
Argyresthiinae adults	small moths
Geometridae	inchworms
Scoliopteryx libatrtix	herald moth
Lymantriidae (unidentified)	tussock moths
Dasychira pudibunda	pale tussock moth
Orgyia antiqua	vapourer moth
Noctuidae (unidentified)	cut worms
Tortricidae (unidentified)	rollers
Tortricid adults	tortrix moths
Araneae	orb-web spiders
Linyphiidae	small spiders
Lycosidae	wolf spiders
Therididae	–
Thomisidae	crab spiders
Opiliones	harvestmen
Psocoptera	booklice
Raphidia	snake flies
Forficula auricularia	earwigs
Molluscs	snails
Trichoptera	caddis flies
Ephemeroptera	mayflies
Isopods	woodlice
Neuroptera	lacewings

Supplement

The following moths were identified by Alastair Burn during two trapping sessions at a site near Lakenheath on 16 June and 6 July 1992

Sphinx ligustri	Privet Hawkmoth
Smerinthus ocellata	Eyed Hawkmoth
Laothoe populi	Poplar Hawkmoth
Pheosia tremula	Swallow Prominent
Pterostoma palpina	Pale Prominent

Spilosoma lubricipeda	White Ermine
Agrotis exclmationis	Heart and Dart
Xestia ditrapezium	Triple-spotted Clay
Lacanobia oleracea	Bright-line Brown-eye
Mythimna impura	Smoky Wainscot
Mythimna pallens	Common Wainscot
Acronicta megacephala	Poplar Grey
Dypterygia scabriuscula	Birdswing
Apamea monoglypha	Dark Arches
Apamea lithoxylaea	Light Arches
Apamea sordens	Rustic Shoulder-knot
Diachrysia chrysitis	Burnished Brass

APPENDIX 4

Site occupancy 1987–2005

	1987	*1988*	*1989*	*1990*	*1991*	*1992*	*1993*	*1994*	*1995*	*1996*	*1997*	*1998*	*1999*	*2000*	*2001*	*2002*	*2003*	*2004*	*2005*
CAM 01	1	1	2	1	1	1	1	2	1	1	1	4	4	4	4	4	4	4	4
CAM 02	1	3	3	2	2	2	3	4	3	3	3	4	4	2	4	4	4	4	4
CAM 03	1	1	4	4	1	4	2		4	2	4	4	4	d					
CAM 04-1	1	1	1	1	1	1	2	2	2	4	4	3	2	4	4	4	4	4	4
CAM 04-2					3	3	3	3	3	4	4	3	2	4	3	4	4	4	4
CAM 04-3						1	1	4	4	4	1	3	4				3	4	4
CAM 04-4							3	3	3	4	4	1	4	4	4	4	4	4	4
CAM 05-1			1	1			3	3	4	3	4	4	4	2	d				
CAM 05-2						3	3	2		3	4	1			3	3	3	4	4
CAM 06			4	4		3	4	4	4	4			4	4				4	4
CAM 07			4	3	4			4					4	d					
CAM 08			4	3	4	3	1	4	4				4	4	4		4	4	4
CAM 10								3		3									
CAM 11													1	2	4	4	4	4	4
SUF 01-1		1	1	1	1	2	1	2	4	3	4	4	4	4	1	4	4	4	2
SUF 01-2									3		4	4	4	4		4	4	4	4
SUF 01-3										4	4		3			4	1	4	2
SUF 02	1	3	4	4	4	4	3	4	3	4	4	4	4	d					
SUF 03		3	4		4	4	3	4			4		4	4	4	4		1	4
SUF 04-1	1	1	1	2	1	2	1	2	1	1	1	1	1	1	3	1	1	1	1
SUF 04-2	1		1										1+1	3	3	3	4	2	3

Key: 1 = confirmed breeding; 2 = probable breeding 3 = bird(s) present; 4 = site visited, no birds recorded

SUF 04-3	1					1	1	1	1	1	1	1	2	1	1	1	1+1	1	1
SUF 04-4	1									1		1	3	4	4	3		4	4
SUF 08								3	4	4	4	4		4					
SUF 09								3	4	4	4	4							
SUF 10								3	4	4	4			4					
NOR 01	1	2	4	4	4	4	4	4	4	4	4		4		4	4	4		
NOR 02-1		1	2	3	3	1	1	3	1	4	3	3	4	4	4	4		4	4
NOR 02-2								3		1	1	3	2	2	4	4		4	4
NOR 03	1	1	1	1	4	1	1	1	3	3	1	1	2	1	1	3		2	4
NOR 04	1	3	1	2	1	3	3	3	3	3	2	4	4	4	3	4	4	4	4
NOR 05	1	1	2	4	4	4	4	4	1	3	4	4	4	4					
NOR 06-1	1	3	1	2	1	1	1	1	4	4	4	4	4	4	4	4		4	4
NOR 06-2						1	1	3	1	1	3	1	4	4	4	4		4	4
NOR 07	1	3	1	2	1	1	1	3	1	4	3	1	4	4	4	4		4	4
NOR 08	1	1	2	3	4	2	2	3	3	2	4	1	3	4	4	4	3		
NOR 09-1	1	1	1	3	4	2	4	2	3	3	3	4	4	4				4	4
NOR 09-2		1	1	3	1	1	3	3	3	3	4	4	4	4				4	4
NOR 09-3			1	1	1	1	2	1	1	2	3	3	4	4	3	4	4	3	4
NOR 10-1	1	1	2	1	1	3	1	3	4	4	4	4	4	4	3	4	3	4	4
NOR 10-2	1	1	4	1	1	1	1	1	3	1	3	3	4	4	4	4	4	4	4
NOR 10-3	1	4	4	4	4	4	4	3	3	4	2	4	4	4	4	4	4	4	4
NOR 11			4		3			4	2	4	1	4	3	3	4	4			
NOR 14	1			3		3		4		1	2				4				
NOR 15			3	1	3	3	2	1	3		1	1	1	1+1	2	1	1+1	2	3
NOR 16			4	4		4	2	4	4							4	4		
NOR 17			2			4	4	4					4	4		4			
NOR 19			1	1	1	4	4	4	4	4	4		4	4	4	4	4		
NOR 20					2		3	4	4	4	4	4	3	4					
NOR 21								1	3	4	3	3	4	4	4	4	4	4	4
NOR 24								4	4	3							4		

	1987	*1988*	*1989*	*1990*	*1991*	*1992*	*1993*	*1994*	*1995*	*1996*	*1997*	*1998*	*1999*	*2000*	*2001*	*2002*	*2003*	*2004*	*2005*
NOR 25								2											
NOR 28									3										
NOR 29									3										
NOR 33										3		3							
NOR 34												1	3	4					
NOR 35																	3		
HUN 01		3	3						4	4									
HUN 02							3		4										
HUN 03							3		4										
HUN 14									3	4	4								
HUN 15											3								

Key: 1 = confirmed breeding; 2 = probable breeding 3 = bird(s) present; 4 = site visited, no birds recorded

APPENDIX 5

'Oriole' in other languages and places

Afrikaans	*europese wielewaal*
Arabic	*sufer*
Azeri	*sarikoinak*
Croatian	*zlatna vuga*
Basque	*urretxori*
Bulgarian	*awliga, ciconco*
Chinese	*yim huangli*
Czech	*zluva obecna*
Danish	*guldpirol*
Dutch	*wielewaal*
Egyptian	*suffayr, abuu suffayr, sufaariya, usfuur at-tuut.*
Estonian	*peoleo*
Finnish	*kuhankeittäjä*
French	*loriot d'Europe*
Georgian	*molaghuri*
German	*pirol*
Greek	σμκοφάγος (*sykophagos*), κιτρινοπομίλι (σμκοφάγος)
Hungarian	*aranymalinko*
Icelandic	*laufglói*
Malay	*peeluk*
Farsi	*parye šahrox* پری شاهرخ طلایی (ج.ش.), مرغ انجیر خوا
Italian	*rigogolo*
Kazakh	*masyk torgai*
Kirghiz	*sary barpy, sargaldak*
Latvian	*vālodze*
Lithuanian	*volung*
Luxembourg	*goldmieral*
Maltese	*taira safra*
Norwegian	*guldtrost*
Polish	*wilga, wywilga*
Portuguese	*papa-figos, verdelhão*
Romanian	*grangural*
Russian	*obyaknovennaya ivolga. lasnaja koshchka,* ИВОЛГА
Serbian	*cichimicha*
Spanish	*oropéndola*
Swedish	*sommargylling*
Turkish	*sarı asma*

APPENDIX 6

Analysis of a recording of an oriole song

Recording made by Chris Knights on 26 May 1995.

Track 1 No oriole song

Track 2 5 minutes 14 seconds

Song phrase	1	weeeow weeow	1
	2	short weee oh	1
	3	weeeoh weeeoh	10 times
	4	eeee oh (short)	1
	5	eee oh (quiet)	1
	1	weeeow weeow	1

ti-ti eeeow eeeow 1
ti-ti eeeow eeeow (second part lower) 1
to eeeow 1

Track 3 6 minutes 28 seconds

1 as above	1
6 as above	17 times with intervals of 3–4 seconds then
4 as above	1 then
6 as above	6 times
7 ooo ooo whoo	1 then
1 as above	9 times
8 wi weeoh	1
9 squark (female?)	1 then
10 eee ow	1 then
6 as above	2 (once faint) then
1 as above	1

Track 4 6 minutes 1 seconds

6 as above	22 times (intervals about the same as above)
1 as above	11 times

Track 5 3 minutes 5 seconds
as above 2
1 as above 8 (moving away as it calls, probably following female)
11 ooh eeeoh (deeper tone) 1
12 weoh (short) 1

Most used song phrase was	6	53 times
Second most used	1	21 times
Total phrases	12	

APPENDIX 7

Details of the 1987 survey

The overall purpose of the survey was to establish the distribution of Golden Orioles in East Anglia, to survey areas where they might colonise in future, to document planting and felling plans by landowners, and to assess what their habitat needs were. We needed to be able to recognise the difference between the poplar hybrid cultivars present, so assistance was sought from the Forestry Commission in the person of their poplar expert John Jobling. He was kind enough to come from his base at Alice Holt, the FC's headquarters in Surrey, and he spent a day with us explaining how we could identify leaf and trunk bark differences.

Each site surveyed was given a name, which was retained on a 'need to know' basis for reasons of confidentiality with respective landowners. They usually only co-operated on this basis, because they understandably did not want other people trespassing in search of what is, in the UK, a rare bird. Each site was given a code, the map reference noted, size and species of tree (if poplar the cultivar also) recorded.

The survey forms for the woodland characteristics took account of the variables as follows:

- poplar cultivar
- % of dead trees
- presence of ditches
- soil conditions
- height of trees
- average circumference of tree trunks
- circumference variation
- average height of branches
- variance of branch heights
- average height of living branches

A further survey form was designed to establish the constituents of understorey, including vegetation (e.g. grazed grass, nettles, umbellifers, bramble, etc.), percentage of bare ground, and the characteristics of the various shrub species, such as alder, elder, sycamore, birch, privet, blackthorn and hawthorn, and species such as snowberry planted for gamebirds in woods. The understorey survey, which was carried out later in the season when plants were more easily identifiable, revealed an immense variety of combinations, which seemed to have no bearing on the suitability or otherwise of a particular site for breeding orioles.

DETAILS OF THE NESTING HISTORY FOR ORIOLES IN 1987

This is described wood by wood but, except at Lakenheath, they are not named for security reasons. Again a standard form was used, detailing as follows:

- reference, name and owner (restricted)
- nest location (within wood)
- nest tree species
- height of tree
- trunk circumference
- branching angle
- height of lowest fork
- height of lowest branch
- height of lowest living branch
- height of nest
- orientation
- distance from trunk (as a proportion of branch)
- presence of artificial fibres in nest
- nest building date (this sometimes can only be approximate according to the state it was in when discovered)
- laying date (again sometimes only approximate because it involved climbing to great heights to be exact and the energy involved, including the time of at least two people, was thought to be inappropriate for the information gained and disturbance to birds at a vulnerable time)
- clutch size (for the same reasons not always accurate)
- hatching date (from ground observation)
- number of nestlings (again from ground except where birds were able to be ringed. Some nests were inaccessible)
- number ringed
- weight and size of pulli
- fledging date
- foraging areas
- food items where possible

The first sections were surveyed by GOG members and consisted of a mixture of plantations and field shelter belts, and sometimes sites included both forms. Details from each were as follows but not in numerical order.

Wood 9 (two pairs bred here in 1985 in separate compartments. In 1986 only one pair bred in an alder)
1987: no evidence of breeding.

Wood 1 (a nest had been found here in 1986)
1987: female not found brooding on eggs or young on 19 June.
Nest height 60ft
Orientation NE
Later found to have failed. Nest retrieved. Lining of greasy black detritus

Wood 6 (pairs seen in here in 1985/6)
1987: none

Wood 10 (pair nested here in 1986/7 using the same tree)
1987: in robusta cultivar
Nest height 40ft
Orientation E
one pullus ringed 23 July, assessed nine days old. Fledging success unknown.

Wood 19 (pair nested 1986/7 using same tree)
1987: pair found nest-building 19 May. Found on 23 June failed.
Nest height 45ft
Orientation NE
Poorly constructed possibly predated.

Wood 20 (pairs nested 1985/6/7)
1987: pair found feeding young 17 June. In eugenii cultivar.
Nest height 36ft
Orientation SE
one pullus ringed 23 June, aged five days. Fledging success unknown.

Wood 21 (field shelter belt, single line (robusta))
1987: nest found 23 June
Nest height 55ft
Orientation SE
Quality very good including red baler twine.
two pulli ringed, aged 8 days.
two watches later found parents foraging 1/4 mile away.

Wood 3 (pair nested in 1986, robusta cultivar)
1987: birds present but no evidence of nests.

Wood 2 (pair nested 1985 and possibly in 1986)
1987: no activity

Wood 16 (pair nested 1985/6, robusta)
1987: nest complete 30 May. Four eggs found 17 June, warm.
22 June nest empty. No detritus. Possibly predated. May have been children because of low nest.

Wood 23 (several shelter belts adjoined at right angles, totalling approximately 400 trees, as measured by the extent the parents were observed foraging in during all this time)
Most studied site. The farm staff report that birds had been present in 1981 also.

History	1982: pair bred, four young fledged	(balsam spire)
	1983: pair bred, two young fledged	(robusta)
	1984: pair bred, three young fledged	(robusta)
	1985: pair bred, three young fledged	(robusta)
	1986: pair bred, three young fledged	(marylandica)

1987: pair again successfully fledged, three young ringed (marylandica)

All the nests were placed an average of 20–25ft from the edge of a fishing pit of about an acre in size.

Wood 17 (birds present 1985/86)
This is an orchard, mainly apples and plums, surrounded by very high poplar shelter belts.
1985: a nest was found only 4ft from the ground in an apple tree (R. Hoblyn, pers. comm.) but no detailed records were kept and GOG did not know at the time.
1986: the nest moved to a poplar belt and placed 40ft high. It was discovered after winter searching, so the outcome is not known.
1987: no evidence of presence or nesting.

Wood 15 (birds bred in 1985/6. Three young ringed in each year. Balsam poplar)
1987: nest found 30 May. Female sitting 10 June and two pulli ringed on 22 June. As there was a considerable size difference between them, age could not be satisfactorily assessed. This badly built nest is thought to have failed later. The female was probably an inexperienced first-year bird because she was very different from the previous year's female, who was an old and very yellow bird.

Wood 18 (birds bred 1985/6; in both cases young were reared in 40ft-high nests)
1987: a nest was found in a different part of this 800-tree site from where nests were placed previously. This nest was not found until 22 June with three eggs in. This late date may indicate that an earlier nest may have been predated and the pair rebuilt their nest away from the original, although that nest was not found subsequently, despite a winter search.

Wood 22 (pair bred in 1985 but discovery was only made by winter nest search. Also bred in 1986, when a family party with one or two young was found during July)
1987: no records because owner denied access to his land.

Wood 14 (pair bred here 1986)
1987: nest discovered 17 June. Two chicks being fed by male at 10-minute intervals.
Nest height 50ft
Orientation SE
On 22 June the male was singing only desultorily three trees away from the nest tree. The female was absent. Three days later the nest was inspected and was found to contain one dead chick. The estimated age at death, according to size of the corpse, was two to three days. Greasy black detritus in floor of nest. If there were other nestlings then they must have been thrown out. Cause of death almost sure to have been pneumonial due to excessive, continuous rain at a vulnerable stage of growth. The male stayed in the area until 30 July but did not defend his territory against intruders.

Details of the 1987 Dagley survey

The following were surveyed by Jez Dagley and are basically the core Lakenheath Bliss Estate birds with others from surrounding land. As he had more time available he monitored how far the birds ranged around sites and noted any overlapping of territories. All of these sites were in block plantations consisting of mainly four types of poplar cultivar – robusta, heidemij, gelrica and serotina (marylandica, balsam spire and TT32 were only found away from this site). They are planted at a rate of 160 per hectare and 8m apart in rows.

Those from the Bliss Estate poplar plantations will be different in so far as they are all connected and form parts of one large block, so there is a greater degree of interaction. However, not all the pairs present on the whole Bliss estate were so closely monitored. Also the historical data cannot be precise, except to say that orioles had used the plantations to a greater or lesser degree since about 1968. Actual territories were not closely monitored in those times, although GOG member Mike Jeanes, with the help of members of the Suffolk Ornithological Group, had carried out singing-bird surveys in some years to assess a potential population. They were able to establish that birds had bred in this area for many years.

The results of the 1987 survey are detailed below.

Main Site Pair 1

(Compartment 79–83) Male arrived 6 May and singing from a serotina cultivar. Sang most days from a similar point until 18 May, when he was joined by a female. It was thought she might have actually arrived on the 16 May, when a large influx of females was noted on the whole estate. Then the female gathered nest material in compartments 85–87, by clinging to bark with feet and from high in the canopy. The male, as is usual, did not take part.

The nest was found on 23 May in compartment 104 (robusta), at a height of 70ft, orientation S., which the pair sometimes returned to over the top of the tree canopy.

Clear felling was already taking place near this site; once the nest was located it had reached only 12 trees away from the nest. It was then deserted and then new nest-building was discovered on 11 June in compartment 82 (heidemij). Height 82ft, orientation E. On 8 July feeding was being undertaken at a rate of every five to 10 minutes. On 13 and 24 July single fledglings were noted being fed by the male, with the female 'catalling' nearby.

Main Site Pair 2

Male arrived 9 May. On 21 May four birds observed chasing each other in the high canopy of compartments 89–92. Only one of them was a female, so obviously there was great competition. From the 22 May only two males were left chasing her, until 26 May. On 30 May she must have decided on her chosen mate and the second male chased off. Collection of nest material was noted in compartments 85–87. On the 18 June the hen bird was incubating the eggs, being serenaded by her mate. On the 22 June they were both by the nest but did not go to it, and next day there were no birds to be seen. However, on the 23 June the male was foraging further away from the nest than usual, in compartments 88–92. Being robusta there was more food in them than in the serotina compartments nearer to the nest. On the 26 June behaviour suggested that hatching had occurred and the female noisily chased off a threatening Cuckoo, after which she returned to feed the young and swallow a faecal sac. On 1 July this pair was seen to fly in over open fields to the south of the plantation, maybe far away to some shelter belts about a mile distant, and after the 8 July they were not seen but everything suggested a successful outcome. The nest here was 22m-high in robusta with orientation E

*Main Site Retained Triangle Pair 3***

Male first noted on 7 May singing from the high canopy. On the 14 May he was seen chasing a female. There occurred on 16 May a large influx of orioles, and so several other males chased after her as well, but by 23 May things settled down and the excess birds departed. On 25 May the nest was found, and on 8 June both birds were at the nest catcalling, believed to be at JD's presence. This indicated they had something to protect, such as eggs. On the 17 May the female was agitated and excited, and so it was assumed that nestlings had hatched, but on 22 May ringers climbed to the nest only to find a dead chick estimated to be six to seven days old. As it was also reckoned to have been dead for about two to three days, this would relate back to the period of prolonged heavy rain referred to previously. The male remained and sang to the end of June but there was no attempt to re-nest. The dead bird was taken for gut analysis later. This nest was 20m-high in heidemij, orientation NE. The nest was also removed to see if this would encourage the pair to rebuild but no luck.

*Main Site Retained Triangle Pair 4***

This male also arrived on the 7 May, singing from the tree high canopy. When the 16 May influx occurred there was a great deal of excitement, as elsewhere, but no female was seen until the 30 May, and on 8 June both birds were catcalling to protest the site. On the 18 June the male chayed at JD's presence, but the female was not on the nest, and on 20 June no birds were seen, only a male singing some way away. Concern was raised as to whether the pair were going to do anything, but on 21 June the hen bird was incubating. With sightings later on 25 and 26 June, pleasure was achieved that all was well. During July the parents were seen at the nest on several dates and on 13 July the pair were feeding a single fledgling. This nest was in robusta, 24m-high, orientation S.

**Authors note. Although JD did not see either of these 'retained triangle' pairs outside the territories outlined in the diagram, it is very possible that they foraged in two nearby places. To the north-west there is a marshy, reedy wetland full of willow trees, known as Botany Bay. To the east, along the Stallode Bank, is a shelter belt consisting of mixed deciduous trees. PM and several other casual observers have noted orioles in these places both before and after 1987, so it is likely they were in one or other of the pairs foraging territories.

APPENDIX 8

Historical records of Golden Oriole in Britain

Birds of Sussex (Walpole Bond)

1865	Blatchington 6/6 fem killed
1891	Fittleworth pair shot (probably bred)
1903	Hamsey nr Lewes pair in residence
1919	Ditchling pair all summer

The Birds of Sussex (Borrer)

Lost date . . .writer claims to have seen 14 in a thorn bush at Henfield Common and 'got within 40–50 yards'

Birds of Surrey (Bucknell)

No confirmed nesting but

1871	Suspected nesting no site details
1890	Pair repeatedly seen four consecutive years near Bellagio

Birds of Surrey (Parr)

1930	Proved breeding near Godstone
1940	Pair attempted to nest Chiswick House (across Thames) but abandoned
1963	Pair laid near Farleigh Common. Eggs robbed

Birds of London Area (London Nat. Hist. Society)

'has bred at least once and has attempted to do so on other occasions'

1940	refers to Chiswick House *see* Surrey

History of the Birds of Hertfordshire (Sage 1959)

1881	Amwell near Ware. Definite breeding nest with three eggs (H. Benningfield and son Trans 11/83)
1886	near Welwyn. Parents shot before young fledged, not authenticated

The Birds of Hertfordshire (Gladwin and Sage)

1970	West of county (Tring) 'closely guarded site and newly fledged young seen' Some doubt has subsequently been cast on this record

The Birds of Essex (Christy)

1887	male 7/6 and two-week period, seen regularly. Believed possible breeding (Baythorn End)
1841	Saffron Walden bred in fir tree. Guarded by landowners. Fledged young

Birds of Oxfordshire (Aplin)
1880 Great Tew. Description of pair shot whilst carrying nesting materials and 'hanging them over the branches'
Birds of Rutland (Hames)
1903 Golden Orioles 'which nested close to Rutland border'
Birds of Northamptonshire and Neighbourhood (Lilford)
'Can hardly be considered very rare in GB between 1860–1865 . . . pair believed to nest near Benefield
Birds of Devon (Durban and Matthew)
1865 'Known to have bred Pilter Abbey, Barnstaple, north Devon'
'It appears regularly each year at Land's End district, Cornwall, where as many as 40 have been seen in a flock'
Birds of Devon (Moore)
1865 Totnes
In Report on Dorset Birds (1951)
1939 Pair. 'Nest long and odd-shaped, suspended between two boughs about 5ft 8ins up in a holly tree. 2 young seen June 10th–30th per FLB'
The Birds of Glamorgan (Cardiff Naturalists Trust)
1863 Penarth. Some evidence pair nested
1883–1886 Coedrhydyglyn. Pair seen and thought to have bred
Birds of Gwent
1967,1969,1970,1974 and1975 Westwood Reservoir, suspected breeding
Birds of Hampshire IOW (Kelsall and Munn)
1884 Egg taken near Christchurch
The Birds of Norfolk (Riviere)
1852 Nr Norwich. Nest in large garden, egg obtained
Birds of Norfolk (Seago)
1955 Horsey Hall. Pair frequented gardens for a month from late May. No evidence of breeding
Birds of Worcestershire (Harthen)
1868 Kyre. Pair seen and probably nested (Prescott)
Birds of Scotland (Thom)
1968 Kinross. June, song and display.
1973 Fife. Singing male.
1974 Fife (same site). Pair seen feeding fledged young. (Letter on record from observer in the Golden Oriole Group files)
Birds of Huntingdonshire and Peterborough (Clark)
1904 Confirmed breeding near Peterborough (Taylor 1958)
1961 Brampton Wood. Pair, believed nested
1985 N. Hunts. Pair, possibly bred (Holme Fen?)
1986 N. Hunts. Pair bred, possible two more (Holme Fen?)
1987 Male present 12 June to 16 July at possible breeding site
The Birds of Kent (Balston, Shepherd and Bartlett)

1836	Ord. Breeding and nest described
1849	Elmstone. Breeding and nest described. Nest taken
1850	Elmstone. Breeding attempted. Female shot
1851	Sandwich. Nest found by gamekeeper and young 'fed to ferrets'.
1869	Female shot and 'on examination I found the bird must have had a nest of eggs'; presumably confirmed by the presence of a brood patch
1874	Dumpton Park. Nest found and protected. Young fledged safely. (J. E. Harting, in *The Field* 3/10/74)
1875	Dumpton Park. Nest and young fledged

Kent Bird Report 1977

1977	East Kent (Blean Wood). Breeding confirmed

Cumbria

Dudden Valley, Cumbria (South Lakeland). Pair breeding in poplars. Trees no longer exist. (Year and reference confidential)

Yorkshire

Near Beswick, Driffield. Orioles bred, nest and young seen. (Year and reference confidential)

Bibliography

Abdulali, H. 1981. *Birds of Sanjay Gandhi National Park*. Bombay Natural History Society, India.

African Bird Club. 2007. Golden Oriole in Haut Niger. *Bulletin of the African Bird Club* 14: 101–102.

Alexander, H. G. 1974. *Seventy Years of Birdwatching*. T & AD Poyser, London.

Ali, S. & Ripley, S. D. 2001. *Handbook of the Birds of India & Pakistan*. Oxford University Press, New Delhi.

Ambedkar, V. C. 1987. Ringing Kundoo. *Journal of the Bombay Natural History Society* 83: 211–212.

Andronova, R. S. 2003. K biologii kitaiskoi ivolgi v Srednem Priamurie [On the biology of the Black-naped Oriole (*Oriolus chinensis*) in the middle Amur River area]. *Ornitologiya* [*Ornithology*] 30: 188–189. In Russian.

Archer, E. & Godman, G. M. 1961. *Birds of British Somaliland and the Gulf of Aden*. Oliver & Boyd, London & Edinburgh.

Artemeyev, A. V., Zimin, V. B., Lapshin, N. V., Sazonov, S. V. & Khokhlova, T. Y. 1993. Order Passeriformes. In *Ornitofauna Karelii* [*Ornithofauna of Karelia*]. Karelian Science Center of Russian Academy of Sciences, Petrozavodsk.

AUA 1999. *A Field Guide to the Birds of Armenia*. American University of Armenia, California.

Aunins, A. 2003. Bird population changes in Latvian farmland. *Latvian Fund for Nature* 12–13: 41–50. Bannerman, D. A. & Bannerman, W. 1965. *Birds of the Atlantic Islands*, Vol 2. Oliver and Boyd, Edinburgh.

Bannerman, D. A. & Lodge, D. 1953. *The Birds of the British Isles*, vol 1. Oliver & Boyd, London.

Barnes, J. A. G. 1960. Golden Oriole breeding in Lancashire. *British Birds* 53: 88–89.

Bau, A. 1907. Goldammer brütet dreimal auf derselben Stelle. *Zeitschrift für Oologie* 17: 24–26.

Baumann, S. 1999. Habitat preferences and home range patterns of Eurasian Golden Oriole in Schleswig Holstein; a radio tracking study. *Corax* 18: 73–87.

Baumann, S. 2000a. Habitat structure and use of Eurasian Golden Orioles *O. oriolus* during breeding and wintering. *Journal of Ornithology* 141: 142–151.

Baumann, S. 2000b. Verhaltensstrategien des Europaeischen Pirols (*Oriolus oriolus*) bei Gesang, Habitatnutzung und Migration [Behavioural strategies of the European Golden Oriole (*Oriolus oriolus*) in respect of song, habitat use and migration]. Unpublished PhD thesis, University of Osnabrueck.

Baumann, S. 2001. Observations on the co-existence of Palearctic and African Orioles (*Oriolus* sp.) in Africa. *Vogelvelte* 122: 67–79.

Beer, W. D. 1961–4. Die Vogelwelt des Braunkohlenbergbaugellaendes im Sueden von Leipzig.

Zoologische Abbhandlungsbereich des Museums der Tierkunde 26: 305–317.
Benson, C. W. & Benson, F. M. 1947. Some breeding and other records from Nyasaland. *Ibis* 89: 279–290.
Benson, S. V. 1970. *Birds of Lebanon.* ICBP, Warne, London.
Bezzel, E., Lechner, F. & Rantfl, H. 1980. *Arbeitsatlas der Brutvoegel Bayerns.* Greven Kilda, Munich.
Bijlsma, R. G. 1995. Wielewalen Oriolus en poularien Populus spec benenden zeeiveau [Golden Orioles and Poplars below sea level]. *Limosa* 68: 21–28.
Bijlsma, R. G. & Hustings, F. 2001. Common and scarce birds in the Netherlands. *Avifauna of the Netherlands* 2: 40–45.
Birfield, I. & van Bommel, F. 2006. *Birds in Europe: population estimates, trend and conservation status.* BirdLife International, Cambridge.
Blinova, T. K. & Blinov, V. N. 1997. *Ptitsy yuzhnogo Zauralya: lesostep i step* [*Birds of southern TransUralia: forest-steppe and steppe*]. Nauka Press (Siberian Branch), Novosibirsk. In Russian.
Blinova, T. K., Mukhacheva, M. M., Kudryavtsev, A. V. & Romanova, N. A. 2001 [*Review of the ornithofauna of the Bolshie Chertany Lake (Tomsk Region)*]. West Siberian Publishing Company, Tomsk. In Russian.
Boddeke, R. 1986. Groepzang van Wielewalen *Oriolus oriolus. Limosa* 59: 96.
Bonham, P. F. & Bonham, M. R. 1971. Studies of less familiar birds 67: Golden Oriole. *British Birds* 64: 385–411.
Borodikhin, I. 1968. *Ptitsy AlmaAty* [*The Birds of AlmaAta*]. Nauka Press (Kazakh SSR), AlmaAta.
Both, C., Bouwhuis, S., Lessels, C. M. & Visser, M. E. 2005. Climate change and population decline in a long distance migratory bird. *Nature* 441: 81–83.
Bouman, M. 1974. *De Wielewaal in Nederland 1966–1973.* Doctoraalverslag 230, Afdeling Natuurbeheer, Landbouwhogeschool, Wageningen.
Brooke, R. K. 1970. Birds eating winged termites. *Honeyguide* 64: 17, 18 & 38.
Brooks, D. J., Evans, M. I., Martins, R. P. & Porter, R. F. 1987. The status of birds: North Yemen and the records of the OSME expedition, Oct–Dec 1985. *Sandgrouse* 9: 4–16.
Brooks, T. M., Pimms, S. L., Kapos, V. & Ravious, C. 1999. Threats from deforestation to montane and lowland birds and mammals in insular South-east Asia. *Journal of Animal Ecology* 68: 1061–1078.
Brosset, A. 1956. Les oiseaux du Maroc oriental de la Méditerranée à Bergent. *Alauda* 24: 161–205.
Burton, J. F. 1995. *Birds and Climate Change.* A&C Black, London.
Buxton, A. 1928. The birds of Geneva. *Ibis* 12 (4): 68–90.
Buxton, A. 1932. Sporting interludes at Geneva. Country Life, London.
Cameron, R. A. D. & Cornwallis, L. 1966. Autumn notes from Azraq, Jordan. *Ibis* 108: 284–287.
Clancey, P. A. 1970. On an oriole—new to the South African list. *Bokmakerie* 22 (3): 53–54.
Clapham, C. S. 1964. Sooty Falcon prey. *Ibis* 106: 376–388.
Clark, J. 1996. *The Birds of Huntingdon and Peterborough.* Cambridge.
Clarke, T., Orgill, C. & Disley, T. 2006. *Birds of the Atlantic Islands.* Christopher Helm Publishers, London.
Coates, B. J. & Bishop, K. D. 1997. *Birds of Wallacea.* Dove Publications, Queensland.
Cody, M. L. (ed.) (1987). *Habitat Selection in Birds.* Academic Press, London.
Collar, N. 1975. Golden Orioles diving into water. *British Birds* 68: 245–246.
Conrad, K. F., Woiwod, I. P., Parsons, M., Fox, R. & Warren, M. S. 2004. Long-term population trends in widespread British moths. *Journal of Insect Conservation* 8: 119–136.
Conrad, K. F., Warren, M. S., Fox, R., Parsons, M. & Woiwod, I. P. 2006. Rapid declines of common, widespread British moths provide evidence of an insect biodiversity crisis. *Biological Conservation* 132: 279–291.
Congreve, W. 1918. Notes from River Somme Valley, Peronne. *Ibis* 10: 348–349.

Coward, T. A. 1920. *Birds of the British Isles and their Eggs*. Warne, London
Cramp, S. & Simmons, C. M. 1993. *The Birds of the Western Palearctic* vol. 7. Oxford University Press, Oxford.
Crette du Palluel, A. 1891. Note sur le Loriot jaune .*Revue des Sciences Naturelles Appliquées* 10: 734–745.
Csiki, E. 1904. Positive Daten über die Nahrung unserer Vögel. *Aquila* 11: 270–313.
Csörgy, T. [also sometimes spelt as Csörgey, T.] 1911. Netz zum Gewollesammeln. *Aquila* 18.
Curry-Lindahl, K. 1981. *Bird Migration in Africa* 1. Academic Press, London Dagley, J. 1987. Golden Oriole Report. Unpublished RSPB report.
Dagley, J. 1994. Golden Orioles in East Anglia and their conservation. *British Birds* 87: 205–219.
Dajovic, M. 1998. Results of bird observations in Belgrade area in 1990–1997. *Ciconia* 7
Davies, N. B. 1992. *Dunnock Behaviour and Social Evolution*. Oxford University Press, Oxford.
DeLucca, C. 1969. *Revised Checklist of the Birds of the Maltese Islands*. Classey, Middlesex.
Dementiev, G. P. & Gladkov, N. A. 1954. *Ptsitsy Sovetskogo Soyusa,* vol 5. Nauka, Moscow.
Devillers, P., Tricot, J., Del Marmol, P., Kerwijn, C. & Anselin, A. 1988. *Atlas van de Belgische broedvogels*. Koninklijk Belgisch Instituut voor Natuurwetenschappen, Brussels.
Diamond, J. 1982. Mimicry of friarbirds by orioles. *Auk* 99 (2): 187–196. Dickinson, E. C. (ed.) 2003a. *The Howard and Moore Complete Checklist of the Birds of the World*. Christopher Helm Publishers, London.
Dickinson, E. C. 2003b. Systematic notes on Asian birds 7. *Zool. Verh. Leiden* 13: 131–146.
Dickinson, E. C. 2004. Systematic notes on Asian birds 42. *Zool. Verh. Leiden* 14: 47–63.
Donatschev, St. 1976. A Polish-ringed oriole recovered in Bulgaria. *Bulletin der Bulgarischen Ornithozentrale Zool. Inst.* 4.
Dorning, H. 1954. Bathing of a Golden Oriole on the top of a nut tree. *Ornithologische Beobachter* 51: 228–229.
Dorzhiev, T. Z. 1996. *Important Bird Areas of the Lake Baikal Region*. Nauka, Irkutsk.
Durand, A. L. 1972. Landbirds over the North Atlantic: unpublished records 1961–65 and thoughts a decade later. *British Birds* 65 (10): 428–441.
Dorzhiev, T. Z. 1996. [*Important bird areas of the Lake Baikal region*]. Eastern Siberian Publishing House, Irkutsk. In Russian.
Dvorak, M. A. & Berg, H.-M. (eds). 1993. *Atlas de Brutvogel Osterreichs*. Umweltbundesamt, Wien.
Dybbro, T. 1976. *De danske ynglefugles udbredelse: Resultaterne af Atlasprojektet, kortlægningen af Danmarks ynglefugle 1971–1974*. Dansk Ornitologisk Forening, København.
Dymond, J. N., Fraser, P. & Gantlett, S. J. M. 1989. *Rare Birds in Britain*. T & AD Poyser, London.
England, M. D. 1964. Females in male plumage. *British Birds* 39: 385, 421.
Ernholdt, T. 1966. Breeding Orioles in Sweden. *Var Fagelvarld* 25: 227–252.
Etienne, P. 1985. Regime Alimentaire du Loriot d'Europe. *Picardie Ecologie* 2 (1): 15–22.
Etienne, P. 1988. *Statut du Loriot dans la Somme et synthèse des données Europeennes sur la reproduction et la migration*. Privately circulated report.
Etienne, P. & Montel, F. 1986. Reproduction en Picardie Du Loriot (*Oriolus oriolus*). *Picardie Ecologie* 2 (1): 73–80.
Evans, C. & Hodder, I. 2006. *Marshland Communities and Cultural Landscape*. Macdonald Institute Monographs, Cambridge.
Fahlberg, G. 2004. Pirolen—en guldfugl synger på sidste vers. *FUGLE i FELTEN* 2. Feather, P. J. 1986. The Bulawayo Garden Survey 1973–1982. *Honeyguide* 32 (1): 13–33.
Fehringer, O. 1922. Die vogelwelt Macedoniens. *Journal für Ornithologie* 70: 89–107, 286–324.
Feige, K.D. 1986. *Der Pirol*. Die Neue Brehm Bucherei, Wittenburg Lutherstadt.
Fenech, N. 1992. *Fatal Flight. The Maltese Obsession with Killing Birds*. Quiller Press, London.
Finlayson, C. 1991. *Birds of the Strait of Gibraltar*. T & AD Poyser, London.

Flade, M. 1994. Quoted in Sutherland, W. J. (2006), *Ecological Census Techniques: a Handbook.* Cambridge University Press, Cambridge.

Fleming, N. 2006. A GIS-based analysis of the special distribution and decline of Golden Orioles in the Fenland Basin of Eastern Britain. Unpublished BSc thesis, University of East Anglia.

Flint, V. E. 1984. *Birds of the USSR.* Princeton University Press, Princeton.

Flint, P. R. & Stewart, P. F. 1983. *The Birds of Cyprus.* British Ornithologists' Union, London.

Foggo, A. 1993. Comparative studies of invertebrate faunas of poplar varieties, and their value as a food resource for the Golden Oriole. Departmental report, Department of Zoology, Oxford.

Formozov, A. N., Osmolovskaya, A. N., Blagosklonov, K. N. (1950). Ptitsy i vrediteli lesa. Znachenie ptic v regulirovanii chislennosti vrednykh nasekomykh lesa i lesnykh posadok [Birds and Forest Pests. Significance of birds in the regulation of pest insects in forests and plantations]. *Sredi prirody [In Nature]* 19: 341–342. In Russian

Fraser, P. A. & Rogers, M. J. 2002. Annual report on scarce migrant birds in Britain. *British Birds* 95: 606–630.

Frith, C. B. 1974. New observations of migrants and vagrants on Aldabra, Farquhar and Astove Atolls, Indian Ocean. *Bulletin of the British Ornithologists' Club* 94: 12–19.

Frith, H. J. (ed.). 1976. *Complete Book of Australian Birds.* Readers Digest, Sydney.

Ganya, I. M. 1969. Vopr ekol prakt snatch ptits mleko. *Moldavii* 4: 26–54.

Garanin, V. I. 1978. *Ptitsy VolzhskoKamskogo Kraya: Vorobinye* [The Birds of the Volga-Kama Territory: Passerines]. Nauka Publishers, Moscow.

Gavrilov, E.I. (1974). Semeistvo Ivolgovye—Oriolidae [Oriole family—Oriolidae]. In M. N. Korelev (ed.), Ptitsy Kazakstana [The Birds of Kazakhstan], volume 5, pp. 5–14. Academiya Nauk (KAZSSR), AlmaAta, Kazakhstan. In Russian.

Gavrilov, E. I. & Gitsov, A. P. 1985. *Sezonnye perelety ptits v predgar'yax Zapadnogo TyaShan.* Publishing House of the Academy of Sciences of the Kazakh SSR, AlmaAta.

Georlette, le M. 1939. Une observation interessante concernant le Loriot jeune. *Gerfaut* 29: 108–110.

Gerasimov, Yu. N., Salnikov, G. M. & Buslaev, S. V. 2000. *Ptitsy Ivanovskoi Oblasti* [Birds of Ivanovo Region]. Rosselkhoz Akademia, Moscow.

Gerrard, V. 2003. *The Story of the Fens.* Robert Hale Ltd, London.

Gilliard, E. T. 1950. Notes on a collection of birds from Bataan, Luzon, Philippine Islands. *Bulletin of the American Museum of Natural History* 94 (8): 302–303.

Ginn, H. B. & Melville, D. S. 1983. *Moult in Birds.* British Trust for Ornithology Guide 19, Thetford, UK.

Ginn, P. J., McIlleron, W. G. & Milstein, P. S. 1989. *The Complete Book of Southern African Birds.* Struik Winchester, Cape Town.

Goodman, S. M. & Meininger, P. L. (eds). 1989. *The Birds of Egypt.* Oxford University Press, Oxford.

Gorman, G. 1994. *Where to Watch Birds in Eastern Europe.* Hamlyn, London.

Graff, H. 1990. First successful breeding of the Golden Oriole by H. Greulich, Reuenberg. *Gefiederte Welt* 114: 363–364.

Graff, H. 1991. Golden Oriole bred again. *Gefiederte Welt.* 115: 365.

Greenoak, F. 1979. *All the Birds of the Air.* Andre Deutch, London.

Grell, M. B. 1998. *Fuglenes Danmark.* Gads Forlag, København.

Grob, J. 1990. Feeding Behaviour. *Nos Oiseaux* 40: 307.

Gromier, E. 1909. Oriole Behaviour. *Revue Francaise d'Ornithologie* 1: 66–72. Gross, W. 1899. Birds copulating on the ground. *Ornithologische Nachrichten* 24: 294–298.

Hagermeijer, E. J. M. & Blair, M. J. (eds). 1997. *ECBB Atlas of European Breeding Birds: Their Distribution and Abundance.* T & AD Poyser, London.

Handrinos, G. & Akriotis, T. 1997. *Birds of Greece.* A&C Black, London.

Hartert, E. 1910. *Die Vogel der palaarktischen Fauna 2.* Berlin.

Hedenström, A. 2003. Aerial hunting by Eleonora's falcons. *Interpretive Birding Bulletin* 4, 2–4.

Herremans, M. 2003. *The study of bird migration across the Western Sahara; a contribution to sound luring.* Royal Museum for Central Africa, Belgium.

Heyder, R. 1952 *Die Voegel des Landes Sachsen.* Leipzig

Holling, M. & the Rare Breeding Birds Panel 2007. Rare breeding birds in the United Kingdom in 2003–2004. *British Birds* 100: 359–360.

Höppner, G. 1954. Courtship behaviour of the Golden Oriole. *Beiträge zur Vogelkunde* 4, 255.

Horner, K. O. 1977. Passage on the Egyptian coast of the Mediterranean. Unpublished PhD thesis, Virginia Polytechnic Institute, Virginia State University.

Hosking, E. 1970. *An Eye for a Bird.* Hutchinson, London.

Houlihan, P. F. & Goodman, S. F. 1986. *The Birds of Ancient Egypt vol 1: The Natural History of Egypt.* Aris and Phillips, Oxford.

Hovorka, W. 1991. Zur Autokologie des Pirols unter Berücksichtigung populationsokologischer Aspekte. Unpublished PhD thesis, University of Vienna.

Hume, A. O. 1889. *The Nests and Eggs of Indian Birds,* vol 1. ReadBooks, UK. Reissued 2006.

Iapichino, C. & Massa, B. 1989. *The Birds of Sicily.* BOU Checklist Series 11. Oxford University Press, Oxford.

Immelmann, K. & Sossinka, R. 1973. Spring migration through Namibia. *Journal Southwest Africa Wissenschaftliche Gesellschaft* 28: 67–71.

Inskipp, T. & Inskipp, C. 1985. *A Guide to the Birds of Nepal.* Croom Helm, London.

Isenmann, P. & Moali, A. 1987. *Oiseaux d'Algerie.* Tour du Valat, France. Ivanov, S. A. & Pogrebnoi, I. P. 2002. [On the bird fauna of Verkhnyaya [Upper] Salda]. In V. K. Ryabitsev (ed.), *Materialy k rasprostraneniyu ptits na Urale, v Priural'ie i Zapadnoi Sibiri* [*Bird distribution in the Urals and West Siberia*], pp. 129. Akademkniga Press, Ekaterinburg. In Russian.

Jaramillo, A. & Burke, P. 1999. *New World Blackbirds: The Icterids.* Christopher Helm Publishers, London.

Jennings, M. C. 1981. *The Birds of Saudi Arabia. A checklist.* Private publication, Whittlesford.

Johansen, H. 1944. Vorlaufiger Bericht uber eine im Sommer 1902 in die Kulundische Steppe und die angrenzenden Teile des Semipalatinsker Gebietes unternommene Reise. *Journal für Ornithologie* 92: 1–105.

Kalyakin, M. V. & Voltzit, O. V. 2005. *Atlas: Birds of Moscow City and the Moscow Region.* Pensoft, Moscow & Sofia.

Kennedy, R. S., Gonzales, P. C., Dickinson, E. C., Miranda, H. C. Jr & Fisher, T. H. 2000. *A Guide to the Birds of the Philippines.* Oxford University Press, Oxford.

Kharkova, O. Y. & Boehme, I. R. 2005. Patterns of location of bird nests in an oak forest in the Nature Reserve 'Les na Vorskle' (Russia). *Berkut* 14: 201–213.

Khokhlov, A. N. 1991. O neobychnom troficheskom povedenii obyknovennoi ivolgi v Stavropol'skom krae [On unusual trophic behaviour of the Golden Oriole in Stavropol Territory]. *Kavkazskiy ornitologicheskiy vestnik* [*Caucasian Ornithological Bulletin*] 2.

King, B. 1984. Immigrant Golden Orioles feeding on the ground. *British Birds* 77: 488.

Kostin, Y. V. 1983. *Ptitsy Krima* [*The Birds of the Crimea*]. Nauka Publishers, Moscow.

Kozlov, N. A. 1988. *Birds of Novosibirsk: a spatial-temporal organisation of the population.* Novosibirsk State Publishing House, Novosibirsk.

Kramer, H. 1980. Bruten Europaischer Vogelarten in Gefangenschaft. *Zoologische Garten Neue Fassung* 89–137. Kullberg, A. 2002. Bird Report: Afghanistan. www.camacdonald.com/birding/

Kumari, E. (ed.) 1952. *Bird Migration.* Academy of Sciences of the Estonian SSR, Tallinn.

Kuresoo, A. & Ader, A. 2000. Point counts of breeding land birds in Estonia. *Hirundo* 1: 200.

Karlavicius, I. P. 2004. Monitoring of breeding birds in Lithuania. *Bird Census News* 13: 77–80.

Labitte, A. 1951. Notes sur la biologie de reproduction d'*Oriolus oriolus* en Pays drouais. *Alauda* 19: 40–48.

Lack, P. 1985. Tsavo East landbird ecology. *Scopus* 9: 23, 57–96.

Lavauden, L. 1924. Voyage de M. Guy Babaut en Tuisie. Resultates scientifique: Oiseaux. Direction Générale de l'Agriculture, du Commerce, et de la Colonisation, Tunis.

Lebreton, P. 1977. *Atlas Ornithologique Rhone-Alpes.* Centre Ornithologique Rhône-Alpes, Université Lyon, Lyon.

Leibak, E., Lilleleht, V. & Veromann, H. 1994. *The Birds of Estonia.* Estonian Academy Publishers, Talinn.

Lewis, A. & Pomeroy, D. 1989. *A Bird Atlas of Kenya.* CRC Press, London.

Lilford, T. L. P. 1883. *Notes on the Birds of Northamptonshire.* Taylor & Francis, London.

Lippens, L. & Wille, H. 1972. *Atlas des Oiseaux de Belgique et d'Europe Occidentale.* Lannoo, Toelt.

Lythe, R. 2006. Golden glow from bird that thinks it's in Africa. *Westmoreland Gazette*, Cumbria.

McGill, L. & Nixon, A. 1991. Garden birds of the Homestead Amanzi 1965–1986. *Beeeater* 42 (1): 39.

Maclean, G. L. 1993. *Roberts Birds of Southern Africa,* vol. 6. Academic Press, London.

MacKinnon, J. 1988. *Birds of Java and Bali.* Gadjah Mada Univ. Press, Jakarta.

MacKinnon, J. & Phillips, K. 1993. *A Field Guide to the Birds of Borneo, Sumatra, Java and Bali.* Oxford University Press, Oxford.

Mal'chevski, A. S. & Pukinski, V. B. 1983. *Ptitsy Leningradski oblasti i sopredel'nykh territoriy 2.* Leningrad State Publishing House, Leningrad. In Russian.

Mallalieu, M. 1988. Birds in Islamabad, Pakistan 1985–87. Unpublished report.

Mason, P. 1993. The Golden Oriole in Cambridgeshire. *CBC annual report 1992*: 61–63.

Meaden, F. 1979. *A Manual of European Bird Keeping.* Blandford, UK.

Meinertzhagen, R. 1920. Quoted in Roberts, T. J. (1991), *Birds of Pakistan.* Oxford University Press, Oxford.

Meinertzhagen, R. 1923. A review of the genus *Oriolus. Ibis* 11 (5): 52–96.

Meinertzhagen, R. 1954. *Birds of Arabia.* Oliver & Boyd, Edinburgh.

Melde, I. & Melde, M. 1977. Zur biologie des pirols. *Falke* 24: 258–263.

Melnikov, Y. I. 1998. Obyknovennaya *Oriolous oriolus* i kitaiskaya *O. chinensis* ivolgi v Vostochnoi Sibiri [Golden Oriole *Oriolus oriolus* and Chinese Oriole *Oriolus chinensis* in Eastern Siberia]. *Russkiy Ornitologicheskiy Zhurnal* [*Russian Journal of Ornithology*] *Expressissue* N45: 17–20. In Russian.

Miller, J. V. 1985. Golden Orioles drinking in flight. *British Birds* 85: 245.

Milwright, R. D. P. 1997. Breeding biology and origin of Golden Orioles in the fenland basin of eastern Britain. *BTO Bird Study* 45: 320–330.

Mooiman, J. 1941. Oriole feeding behaviour. *Ostrich* 12: 11–16.

Moreau, R.E. 1961. The palearctic bird migration systems. *Ibis* 103a: 373–427.

Mukhtarov, H.S. 1991. Akusticheskoe i territorialnoe povedenie obyknovennoi ivolgi v period gnezdovaniya [Acoustic and territorial behaviour of the Golden Oriole during the nesting period]. In A. N. Khokhlov (ed.), *Fauna, naselenie i ekologiya ptits Severnogo Kavkaza* [*Fauna, population and ecology of the birds of northern Caucasia*], pp. 26–28. Stavropol Pedagogical University Press, Stavropol. In Russian.

Mwangi, M. A. K. 2006. Conservation status of Eurasian Golden Oriole habitat in the Fenlands of East Anglia, UK. Unpublished B.Sc dissertation, University of East Anglia.

Natorp, O. 1938. Zur brutbiologie des pirols (*Oriolus oriolus*). *Beitrag des Fortfluges (Vogel)* 14: 12–13.

Neff, G. 2003. The Golden Oriole . . . an exotic bird in our avifauna. *Gefiederte Welt* 127: 326–329, 366–368.

Nicholson, E. M. 1950. *Birds and Men.* Collins, London.

Nightingale, B. & Dempsey, E. 2007. Recent reports. *British Birds* 100: 454.

Nikolaev, V. I. 1998. *Ptitsy bolotnykh landshaftov natsional'nogo parka 'Zavidovo' i Verkhnevolzhya* [Birds of the marshes of the 'Zavidovo' national park and Upper Volga area]. State complex 'Zavidovo', Tver. Norman, D. 1994. *The Fieldfare.* Hamlyn, London.

Noskov, G. A., Zimin, V. B., Rezvyi, S. P., Rymkevich, T. A., Lapshin, N. V. & Golovan, V. I. 1981. *Ptitsy Ladozjskogo orn. stationara i ego okrestonstei* [*The Birds of Ladoga Lake ornithological permanent station and its vicinities*]. Leningrad State University Press, Leningrad.

Oatley, T. B. 1969. Bird ecology of the evergreen forests of north western Zambia. *Puku* 5: 141–180.

Paulussen, W. 1955. *Oriolus oriolus* (L). *Gerfaut* 45: 69–71.

Paulussen, W. 1993. *De Broedvogels van de Turnhoutse Kempen* (*The Breeding Birds of the Turnhout Campine*) *1942–1992*. Oriel Stringer, Amsterdam.

Pazzuconi, A. 1997. *Uova e Nidi degli Uccelli d'Italia*. Edizioni Calderini, Bologna.

Pedersen, F. 1972. The breeding population of Golden Oriole in Denmark. *Feltornithologen* 14: 20–23.

Pershakov, A. A. 1932. Cited in Garanin, V. I. (1978), *Ptitsy Volzhsko-Kamskogo kraya. Vorobinye* [*The Birds of the Volga-Kama territory. Passerines*]. Nauka Publishers, Moscow. In Russian.

Petrushenko, A. A. & Klestov, N. L. 1991. *O pitanii ivolgi v Ukrainskom* [*The Golden Oriole in Ukraine*]. Nauka i Technika, Minsk

Popov, V. A. 1978. *Ptitsy Volzhsko-Kamskogo kraya* [*Birds of the Volga-Kama Territory*]. Vol. 2: Passeriformes. Nauka Publishers, Moscow. In Russian.

Prokop, P. 1983. Golden Orioles feeding on the ground. *British Birds* 75: 185.

Rasmussen, P. C. & Anderton, J. C. 2005. *Birds of South Asia* vol. 1. Lynx Edicions, Barcelona.

RBBP. Annual reports of the Rare Breeding Birds Panel. British Birds, St Leonards on Sea.

Reinsch, A. 1958. Am neste des Pirols. *Vogelwelt* 79: 154–157.

Reinsch, A. 1959. Beobachtungen am neste des Pirols. *Vogelwelt* 80: 149–156.

Reinsch, A. & Warncke, K. 1971. Zur Brutbiologie des Pirols. *Vogelwelt* 92: 121–141.

Rheinwald, G. 1982. *Brutvogelatlas der Bundesrepublik Deutschlands*. Schriftenreihe des Dachverband, Bonn.

Ristow. D, Wink. C, & Wink. M. 1984. Asscssment of Mediterranean Autumn migration by Prey Analysis of Eleonora's Falcon. *Alle Ricerche Biologhiche della Salvaggina 10: Proceedings of the 1st conference on birds wintering in the Mediterranean Region, Aulla, February 1984*, pp. 285–295 (supplement).

Roberts, A. 1928. Birds and mammals from Southwest Africa. *Annals of the Transvaal Museum* 12: 289–329.

Roberts, E. L. 1954. *The Birds of Malta*. Progress Press, Malta.

Roberts, T. J. 1991. *Birds of Pakistan*. Oxford University Press, Oxford.

Robertson, I. S. 1980. Migrant Golden Oriole feeding on ground. *British Birds* 73: 265.

Rudov, F. 1887. Was geschieht mit den alten Vogelnestern? *Ornithologische Monatsschrift* 12: 351–357.

Ruthke, P. 1951. Die Brutvögel des Mönnegebietes im pommerschen Oberdelta. *Orn. Abh.* 1: 4–20.

Ryabitsev, V. K., Tarasov, V. V., Primak, I. V., Polyakov, V. E., Grekhov, R. G. & Bologov, I. O. 2002. [On the bird fauna of the south of Kurgan Region]. In V. K. Ryabitsev (ed.), *Materialy k rasprostraneniyu ptits na Urale, v Priural'ie i Zapadnoi Sibiri* [Bird distribution in the Urals and West Siberia], p. 211. Akademkniga Press, Ekaterinburg. In Russian.

Rydzewski, W. 1978. The longevity of ringed birds. *Ring* 967: 218–262.

Sagitov, A. K. 1965. Gnezdovaya zhizn ivolgi [Nesting by Orioles]. In M. N. Korelov & I. A. Dolgushin (eds), *Novosti ornitologii* [*Ornithology News*], pp. 333–334. Nauka of the Kazakh SSR, Alma-Ata. In Russian.

Sagitov, A. K. & Bakaev, S. B. 1980. *Ekologiya gnezdovaniya massovykh vidov ptits yugo-zapadnogo Uzbekistana* [*Breeding ecology of the birds of south-western Uzbekistan*]. FAN Publishers, Tashkent. In Russian.

Salvan, J. 1966a. Spring arrival and nesting in 1966. *Nos Oiseaux* 24: 64.

Salvan, J. (1966b) Prey taken by raptors, 1959–66. *Nos Oiseaux* 24: 267.

Sanderson, F. J., Donald, P. F., Pain, D. J., Burfield, I. J. & van Bommel, F. P. J. 2006. Longterm population declines in Afro-Palearctic migrant birds. *Biological Conservation* 131: 93–105.

Schiffer, H. 2005. Breeding of *O. chinensis* in captivity. *Avicultural Magazine* l3 (3): 144.

Serra, G., Al-Qaim, G., Abdallah, M. S., Kanani, A. & Assaed, A. K. 2005. Migrant passage through Syria. *Sandgrouse* 27: 104–125.
Shaposhnikov, L. 1938. Cited in Garanin, V. I. (1978), *Ptitsy Volzhsko-Kamskogo kraya: Vorobinye* [*The Birds of the Volga-Kama Territory. Passerines*]. Nauka Publishers, Moscow.
Shergalin, J. Birds in Russia website. http://groups.yahoo.com/group/BirdsinRussia
Shilova, N. V. 1953. Cited in Garanin, V. I. (1978), *Ptitsy Volzhsko Kamskogo Kraya: Vorobinye* [The Birds of the Volga-Kama Territory: Passerines]. Nauka Publishers, Moscow.
Shirihai, H. 1996. *The Birds of Israel.* Princeton University Press, Princeton.
Shnitnikov, V. M. 1949. *Birds of the Semirichie.* Nauka i Technika, Moscow & Leningrad.
Sibley, B. L. & Monroe, C. G., Jr. 1993. *A World Checklist of Birds.* Yale University Press, New Haven.
Skead, C. J. 1995. *Life History Notes on East Cape Bird Species 1940–1990*, vol 1. Algoa Regional Services Council, Port Elizabeth.
Snell, H. 1994. Opportunities for increasing farm income from poplar. Paper presented to RASE conference 'Poplar—the Golden Opportunity', 24 October 1994, Stoneleigh Park, Warwickshire.
Sonnenschein, E. 1998. Totfund eines Pirols (*Oriolus oriolus*) am Kilimanjaro (Tansania), in 4.300 m ü M. *Die Vogelwarte* 39: 300–301.
SOVON. 1987. *Atlas der Nederland Vogels.* Arnhem Samenwerkender Organisaties Vogel-Onderzoek, Arnhem.
Stahlbaum, G. 1966. Oriole communal bathing. *Ornithologische Mittteilungen* 18: 141.
Stakheev, V. A. 2000. *Ptitsy Altaiskogo zapovednika* [*Birds of the Altai Nature Reserve*]. Association of Yenisei Nature Reserves and National Parks, Shushenskoe. In Russian.
Storrs, J. 1994. Poplar: the material, its potential uses and markets. Paper presented to RASE conference 'Poplar—the Golden Opportunity', 24 October 1994, Stoneleigh Park, Warwickshire.
Strange, M. 2001. *Photographic Guide to the Birds of Indonesia.* Christopher Helm, London.
Stresemann, E. 1948. Die Wanderungen des Pirols. *Ornithologische Beobachter* 1: 126–142.
Sultanov, E. G. 1988. Organizatsiya pesni u ivolgi obyknovennoi (*Oriolus oriolus* L.) [Song organization (structure) in the Golden Oriole]. *Izvestiya Akademii Nauk Azerbaidjanskoi SSR* [*Proceedings of the Academy of Sciences of Azerbaijan SSR, Series Biological Sciences*] N1: 57–62.
Svensson, L. 1992. *Identification Guide to European Passerines*, 4th edition. Published by the author, Stockholm.
Sylven, M. 1978. Sommargyllingens förekomst i Skäne, 1977 [The occurrence of the Golden Oriole in Skäne, south Sweden, 1977]. *Anser* 17: 29–34.
Szep, T. & Nagy, K. 2002. *Mindennapi Madaraink Monitoingia 1999–2000.* MME Birdlife Hungary, Budapest.
Tabbush, P. & Beaton, A. 1998. Hybrid poplars: present status and potential in Britain. *Forestry* 71 (4): 355–364.
Tarasov, V. V., Ryabitsev, V. K., Grekhov, R. G., Lyakhov, A. G. , Bologov, I. O. 2002. [On the bird fauna of the south of Kurgan Region]. In V. K. Ryabitsev (ed.), *Materialy k rasprostraneniyu ptits na Urale, v Priural'ie i Zapadnoi Sibiri* [*Bird distribution in the Urals and West Siberia*], p. 211. Akademkniga Press, Ekaterinburg. In Russian.
Tarboton, W. R., Kemp, M. I. & Kemp, A. C. 1987. *Birds of Transvaal.* Transvaal Museum, Pretoria.
Teixera, R. M. (ed.) 1979. *Atlas van de Nederlandse broedvogels.* Deventer, Amsterdam.
Thevenot, M., Vernon, R. & Bergier, P. 2004. *Birds of Morocco.* British Ornithologists' Union, London.
Tiainen, J. 1979. Kuhkankeittajat (Golden Oriole). In O. Hilden, J. Tiainen & R. Valjaka (eds), *Muultolinnut*, pp. 264–271. BirdLife Finland, Helsinki.
Tsarev, A. P. 2006. Natural poplar and willow ecosystems on a grand scale in the Russian federation. Food and Agriculture Organisation of the United Nations online: www.fao.org.

Tsybulin, S. M. 1999. *Ptitsy Severnogo Altaya* [*Birds of the Northern Altai*]. Nauka Publishers, Novosibirsk.

Tucker, L. 1989. In Ogilvie, M. & Winter, S. (eds), *Best Days with British Birds*. British Birds, Blunham, Bedfordshire.

Uhlenhaut, K. 1998. Ethologische Beobachtungen am Brutplatz des Pirols. *Ornithologische Journal Museum Heinemann* 16: 65–84.

Uhlenhaut, K. 1999. Observations and investigations of breeding biology of Golden Oriole (*Oriolus oriolus*). *Ornithologische Journal Museum Heinemann* 17: 1–91.

van den Berg, A. & Lafontaine, D. 1995. *Where to Watch Birds in Holland, Belgium & Northern France*. Hamlyn, London.

Ushakov, S. L. 1927. Spisok ptits Permskogo okruga Ural'skoi oblasti [Birds of the Perm district of Ural Region]. [*Bulletin of Moscow Naturalists' Society Department of Biology, New series*] 36: 68–116. In Russian.

van der Linde, K. 1996. A further record of the Isabela Oriole. *Forktail* 11: 171.

van Someren, V. G. L. 1956. *Days with Birds: Studies of habits of some East African species. Fieldiana: Zoology* 38. Chicago Natural History Museum, Chicago.

Vaurie, C. 1972. *Tibet and its Birds*. Witherby, London.

Vernon, C. J. 1984. Population dynamics of birds in Brachystegia woodland. In J. A. Ledger (ed.), *Proceedings of the 5th Pan-Africa Ornithological Congress*, pp. 201–216. Southern African Ornithological Society, Johannesburg.

Vietinghoff, A. V. 1927. Ernahrungsbiologie und soziale Strukture. *Mitteilungen, Verein für Sachsene Ornithologie* 2: 81–93, 133–148.

Vilbaste, H. 1990. Number dynamics of the breeding birds in the forests of south-west Estonia. In O. Renno (ed.), *Communications of the Baltic Commission for the Study of Bird Migration No 22*, pp.102–117. Tartu, Riga, Estonia.

Von Knorre, D., Grün, G., Günther, R. & Schmidt, K. (eds). 1986. *Die Vogelwelt Thüringens*, vol 3. Avifauna der DDR, Jena.

von Richter, G. 1949. Mit der camera am Pirolnest. *Vogelwelt* 2: 336.

Waring, P., Townsend, M. & Lewington, R. 2003. *Field Guide to the Moths of Great Britain and Ireland*. British Wildlife, Hampshire.

Wassmann, R. 1986. Zur Nahrung des Pirols (*Oriolus oriolus*). *Vogelwelt* 107: 156.

Wassmann, R. 1990. Der Pirol Zur biologie des Vogels des Jahres 1990. *Berlin ANL* 14: 153–160.

Wassmann, R. 2004. *Der Pirol: ein Tropenwaldvogel in Europa?* AULAVerlag, Wiebelsheim.

Weerd, M. J. & Hutchinson, R. 2005. Recent observations of the Isabela Oriole in the Philippines. *Forktail* 20: 133–138

Whitaker, J. I. S. 1905. *Birds of Tunisia*. Witherby, London.

Winkler, R. 1999. Avifaune des Suisse. *Nos Oiseaux* occasional supplement 3: 175.

Yeatman, L. (ed.) 1976. *Atlas des oiseaux nicheurs de France*. Société Ornithologique de France, Paris.

Yeatman-Berthelot, D. & Jarry, G. 1994. *Nouvel atlas des oiseaux nicheurs de France 1985–1989*. Société Ornithologique de France, Paris.

Zarth, K. 1968. Zum alter von Käfigvögeln. *Gefiederte Welt* 92: 140.

Zarudni, N. 1888. Cited in Osmolovskaya, V. I. & Formozov, A. N. (1950), *Ocherki ekologii nekotorykh poleznykh ptits lesa* [*Ecological sketches of some useful forest birds*]. Nauka Publishers, Moscow. In Russian.

Zippelius, H. M. 1972. Zur Brutbiologie des Pirols. *Bonner Zoologische Beitraege*. 23: 338–346.

Index

Page numbers in italics refer to illustrations and tables.